Dr. Clinton A. Johnson
PRACTICE OF CHIROPRACTIC

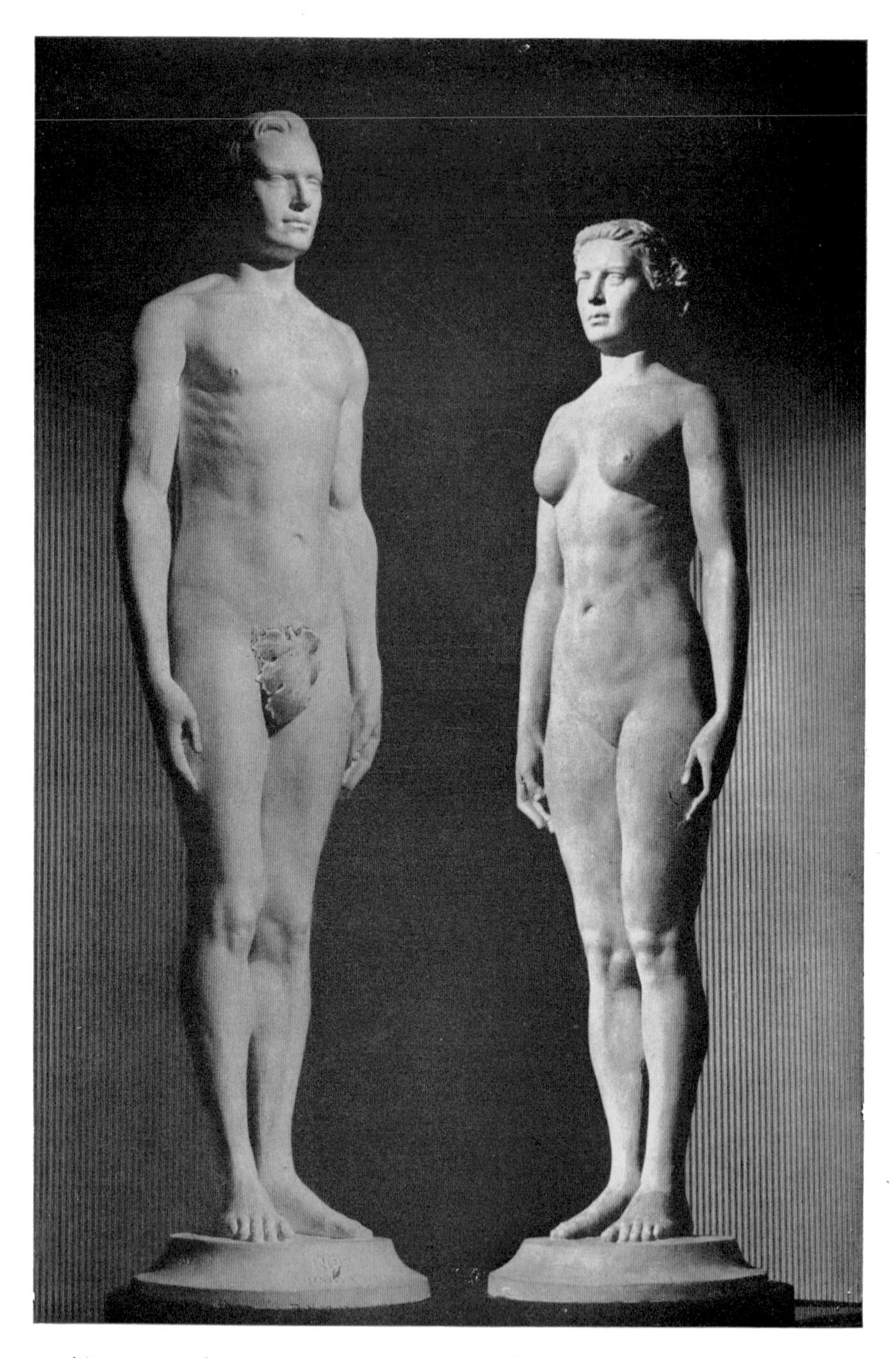

NORMMAN—Average American Male, age 20, height 68½ inches, weight 155 pounds.

NORMA—Average American Woman, age 20-35, height 63½ inches, weight 121-127 pounds.

Courtesy of the Metropolitan Life Insurance Company

THE HEAD, NECK, AND TRUNK

MUSCLES AND MOTOR POINTS

BY

JOHN H. WARFEL, Ph.D.

ASSOCIATE PROFESSOR OF ANATOMY, STATE UNIVERSITY OF NEW YORK AT BUFFALO
SCHOOL OF MEDICINE, BUFFALO, NEW YORK

Fourth Edition

111 Illustrations

LEA & FEBIGER

PHILADELPHIA

First Edition 1947

Reprinted
May, 1949
November, 1951
August, 1953
February, 1955
October, 1956
February, 1958

Second Edition, 1960
Reprinted
August, 1963
June, 1965

Third Edition, 1967
Reprinted November, 1969

Fourth Edition, 1973

Reprinted
April, 1976

Warfel, John H.
The head, neck, and trunk.

Previous editions by D. P. Quiring.
1. Anatomy, Human. 2. Anatomy, Surgical and topographical. I. Quiring, Daniel Paul, 1894-1958. The head, neck, and trunk. II. Title. [DNLM: 1. Anatomy, Regional—Atlases. 2. Muscles—Anatomy and histology—Atlases. WE 17 W274h 1973]
QM531.W37 1973 611'.73 73-5866
ISBN 0-8121-0453-6

ISBN 0-8121-0453-6

Library of Congress Catalog Card Number 73-5866

Published in Great Britain by Henry Kimpton Publishers, London

Printed in the United States of America

PREFACE

THE FIRST EDITION of The Head, Neck and Trunk was prepared by Daniel P. Quiring of Cleveland, Ohio, to portray in diagrams and condensed descriptions the individual muscles of the head, neck, and trunk, together with their chief arterial and nerve supply.

The muscles of the left side have been shown throughout. Their Latin names according to Gray's Anatomy have been employed. Since the trunk and deeper neck muscles do not lend themselves to accurate electrical testing, only the general motor points of the face and neck are shown in a single figure.

In this fourth edition reference corrections have been made to the 29th American Edition of Gray's Anatomy (Philadelphia, Lea & Febiger, 1973), and two of the original plates have been redrawn.

The major change has been in the elimination of the references to Cunningham's Textbook of Anatomy. In their place references are now given to the 6th edition of Grant's Atlas of Anatomy (Baltimore, The Williams & Wilkins Co., 1972). The reader will notice that the references to Grant for the nerve and/or arterial supply to some of the muscles presented in this book are indicated as "not shown." This was done when it was found that the atlas did not have plates showing the specific relationship between the given muscle and its nerve or artery.

It is hoped that the reference combination between a standard textbook of anatomy and a popular atlas will render this book of more service to the user.

Buffalo, New York — John H. Warfel

CONTENTS

MUSCLES OF THE SCALP AND FACE

Epicranius (Occipitofrontalis) 12
Temporoparietalis 13
Auriculares 14
Orbicularis oculi 15
Corrugator supercilii 16
Procerus (Pyramidalis nasi) 17
Compressor naris (Nasalis) 18
Dilatator naris (Nasalis) 19
Depressor septi 20
Levator labii superioris 21
Levator anguli oris (Caninus) 22
Zygomaticus (major) 23
Risorius 24
Depressor (Quadratus) labii inferioris 25
Depressor anguli oris (Triangularis) 26
Mentalis 27
Orbicularis oris 28
Buccinator 29

MUSCLES OF MASTICATION

Temporalis 30
Masseter 31
Pterygoideus medialis (internus) 32
Pterygoideus lateralis (externus) 33

MUSCLES OF THE NECK

a) Cervical
Platysma 34
Sternocleidomastoideus 35
b) Suprahyoid
Digastricus 36
Stylohyoideus 37
Mylohyoideus 38
Geniohyoideus 39

c) Infrahyoid
Sternohyoideus 40
Sternothyreoideus 41
Thyreohyoideus 42
Omohyoideus 43
d) Anterior vertebral
Longus colli (L. cervicis) 44
Longus capitis 45
Rectus capitis anterior 46
Rectus capitis lateralis 47
e) Lateral vertebral
Scalenus anterior 48
Scalenus medius. 49
Scalenus posterior 50

DEEP MUSCLES OF THE BACK

Splenius capitis and cervicis 51
Erector spinae (Sacrospinalis). 52
Iliocostalis lumborum, I. thoracis, I. cervicis 53
Longissimus thoracis, L. cervicis, L. capitis 54
Spinalis thoracis, S. cervicis, S. capitis 55
Semispinalis thoracis, Ss. cervicis, Ss. capitis. . . . 56
Multifidus 57
Rotatores 58
Interspinales 59
Intertransversarii 60

SUBOCCIPITAL MUSCLES

Rectus capitis posterior major 61
Rectus capitis posterior minor 62
Obliquus capitis inferior 63
Obliquus capitis superior 64

MUSCLES OF THE THORAX

Intercostales externi 65
Intercostales interni 66
Subcostales 67
Transversus thoracis 68
Levatores costarum 69
Serratus posterior superior. 70
Serratus posterior inferior 71
Diaphragm 72

CONTENTS

*MUSCLES OF THE ABDOMEN

Obliquus externus abdominis 73
Obliquus internus abdominis 74
Cremaster 75
Transversus abdominis 76
Rectus abdominis 77
Pyramidalis 78
Quadratus lumborum 79

MUSCLES OF THE PELVIS

Levator ani 80
Coccygeus (Ischio-coccygeus) 81

MUSCLES OF THE PERINEUM

Transversus perinei superficialis 82
Bulbocavernosus (male) 83
Bulbocavernosus (female) 84
Ischiocavernosus 85
Transversus perinei profundus (male) 86
Transversus perinei profundus (female) 87
Sphincter urethrae 88
Sphincter ani externus (and S. ani internus) 89

MUSCLES OF THE EYE

Levator palpebrae superioris 90
Recti: superior and inferior 91
Recti: lateralis and medialis 92
Obliquus superior 93
Obliquus inferior 94

MUSCLES OF THE TYMPANIC CAVITY

Tensor tympani 95
Stapedius 96

MUSCLES OF THE LARYNX (Intrinsic)

Cricothyreoideus 97
Cricoarytenoideus posterior 98
Cricoarytenoideus lateralis 99
Arytenoideus: obliquus and transversus 100
Thyreoarytenoideus 101
Vocalis 102
Thyreoepiglotticus 103

*Posterior abdominal muscles, including psoas major and minor, and iliacus, are described in The Extremities.

CONTENTS

MUSCLES OF THE TONGUE

(Extrinsic)
Genioglossus 104
Hyoglossus (and Chondroglossus) 105
Styloglossus 106
(Intrinsic)
Longitudinalis (linguae superior and inferior) . . 107
Transversus and verticalis linguae 108

MUSCLES OF THE PALATE

Levator veli palatini (Levator palati) 109
Tensor veli palatini (Tensor palati) 110
Musculus uvulae 111
Palatoglossus (glossopalatinus) 112
Palatopharyngeus (pharyngopalatinus) 113

MUSCLES OF THE PHARYNX

Constrictor pharyngis inferior 114
Constrictor pharyngis medius 115
Constrictor pharyngis superior 116
Stylopharyngeus 117
Salpingopharyngeus 118
Motor Points 119

CHARTS

Foramina of the skull (norma basalis) 120–121
Foramina of the skull (base, interior) 122–123

INDEX 125

EPICRANIUS (Occipito-frontalis)

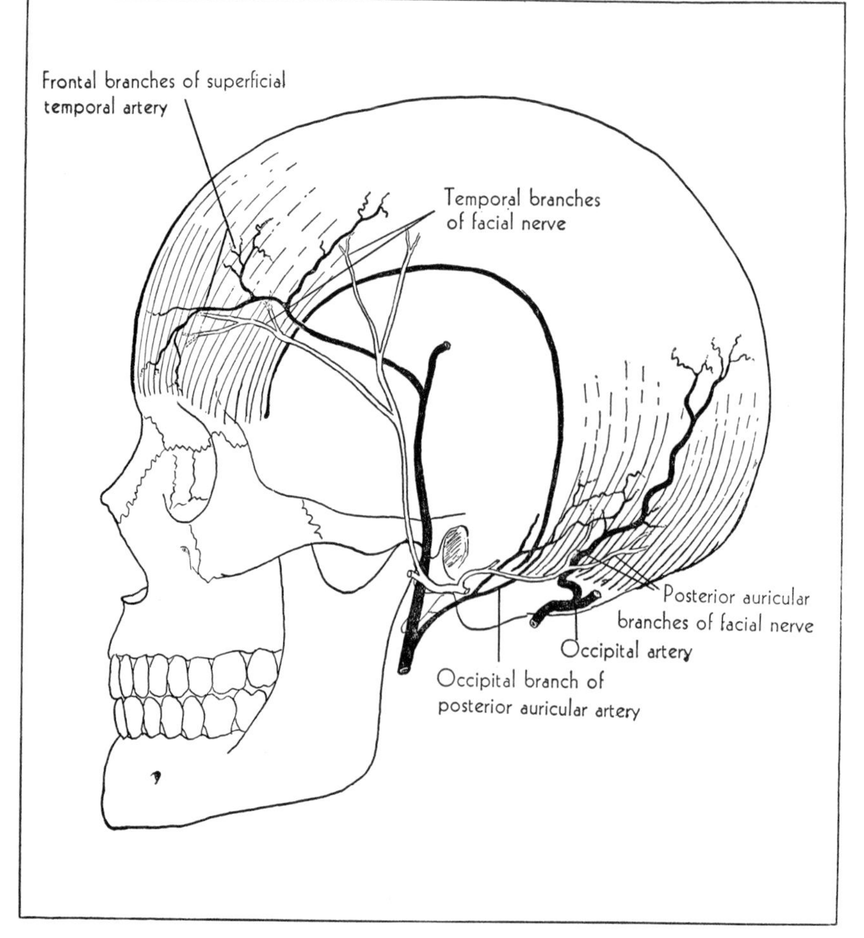

ORIGIN: Occipital bellies from lateral two-thirds of superior nuchal line and mastoid process, frontal bellies from epicranial aponeurosis at level of coronal suture

INSERTION: Skin of occipital region, skin of frontal region and galea aponeurotica

FUNCTION: Moves scalp backward and forward, raises eyebrows (surprise)

NERVE: Temporal branches of facial supply frontal belly, posterior auricular branches of facial supply occipital belly

ARTERY: Frontal branch of superficial temporal to frontal belly, occipital branch of posterior auricular and descending branch of occipital to occipital belly

REFERENCES:

	GRAY	GRANT'S ATLAS
Muscle	379	466, 490
Nerve	380, 924, 930, 931	550, 552
Artery	585, 586, 587	24, 466, 467, 551

TEMPOROPARIETALIS

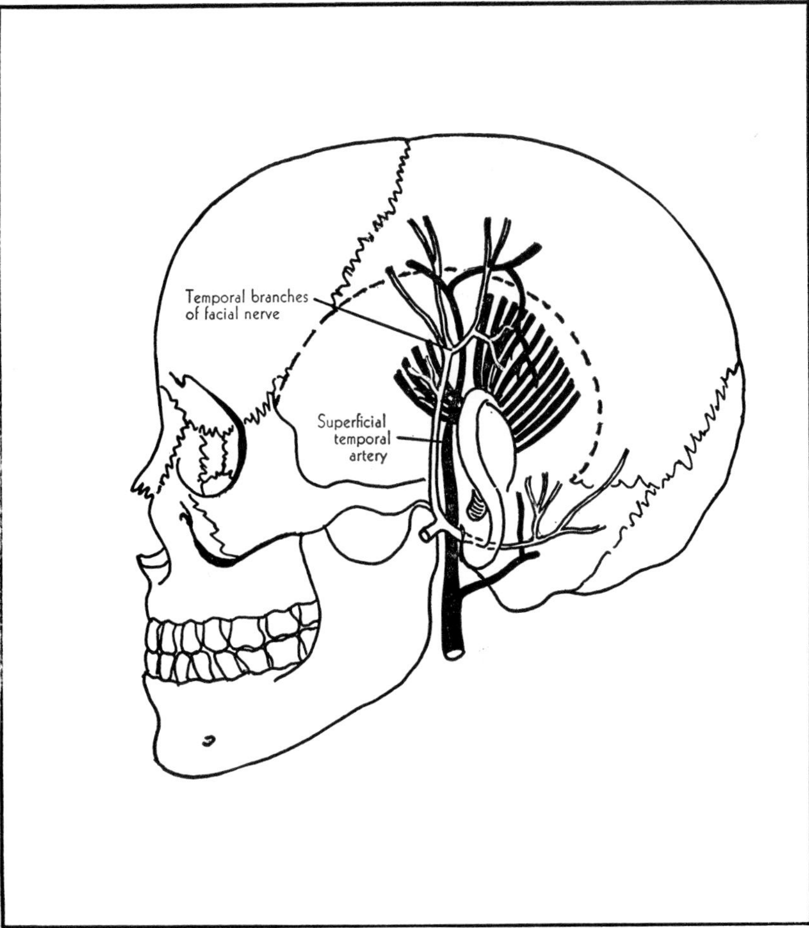

ORIGIN: Temporal fascia above and anterior to the ear
INSERTION: Lateral border of the galea aponeurotica
FUNCTION: Tightens scalp, draws back skin of temples
NERVE: Temporal branches of facial
ARTERY: Auricular, temporal and parietal branches of superficial temporal

REFERENCES: GRAY
- Muscle 380
- Nerve 380, 924, 931
- Artery 586

GRANT'S ATLAS
- Not listed in 6th edition

This muscle, newly designated in the PNA, supplants the thin muscular sheet formerly designated as the Auricularis superior and Auricularis anterior

AURICULARES (anterior, superior, posterior)

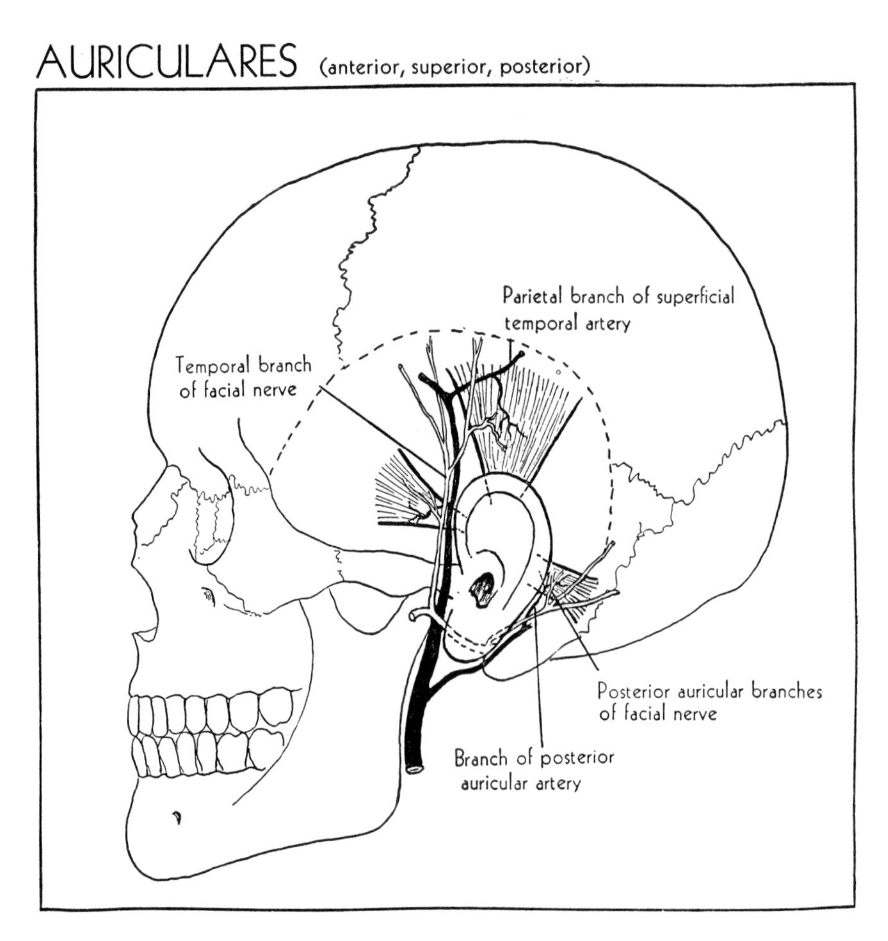

ORIGIN: a) A. anterior from temporal fascia and epicranial aponeurosis
b) A. superior from epicranial aponeurosis and temporal fascia
c) A. posterior from surface of mastoid process

INSERTION: a) A. anterior into anterior part of medial surface of helix
b) A. superior into upper part of medial surface of auricle
c) A. posterior into lower part of cranial surface of auricle

FUNCTION: Retract and elevate ear, usually rudimentary and functionless

NERVE: Posterior auricular and temporal branches of facial

ARTERY: Auricular branch of posterior auricular, parietal branch of superficial temporal

REFERENCES:	GRAY	GRANT'S ATLAS
Muscle	380	467, 472
Nerve	381, 924, 930, 931	550
Artery	585-586	466, 467, 550

These are small, inconstant muscle bellies closely associated with the auricula

ORBICULARIS OCULI

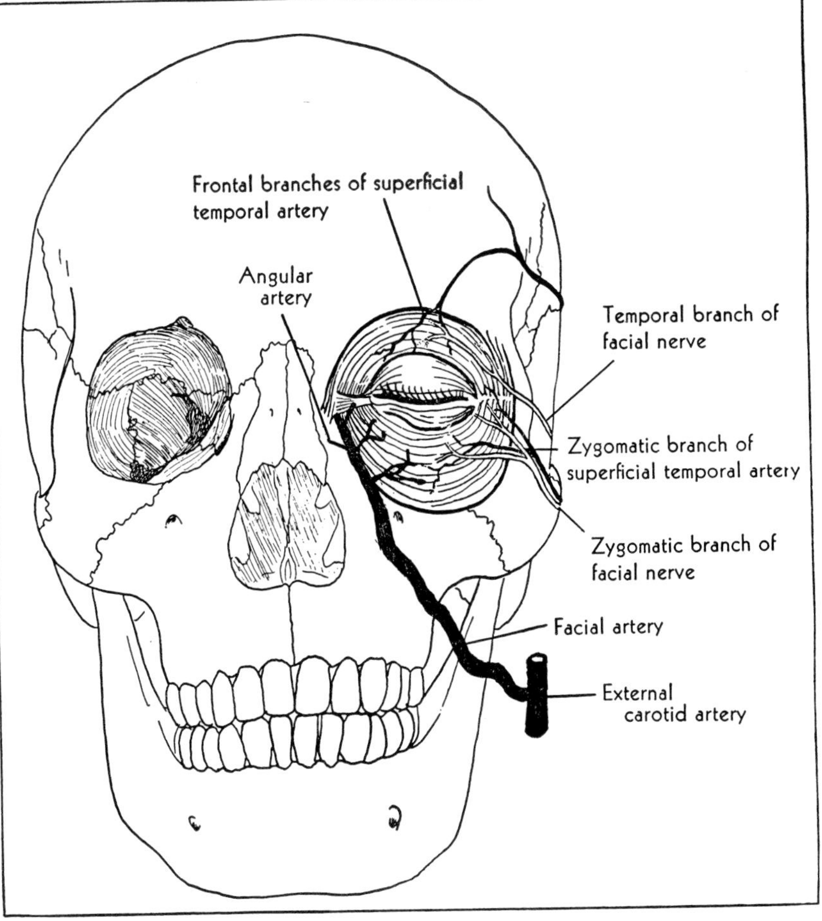

ORIGIN: a) Orbital part from medial orbital margin; b) Palpebral part from palpebral ligament; c) Lacrimal part from lacrimal bone

INSERTION: a) Orbital fibers arch around upper lid and return around lower to palpebral ligament; b) Palpebral fibers interlace at lateral angle of eye in palpebral raphé; c) Lacrimal part into medial portion of upper and lower eyelids

FUNCTION: Sphincter of eyelids; palpebral part involuntary

NERVE: Temporal and zygomatic branches of facial

ARTERY: Zygomatic and frontal branches of superficial temporal, angular branch of facial

REFERENCES:	GRAY	GRANT'S ATLAS
Muscle	381	466, 467
Nerve	382, 924, 931	467
Artery	584, 586	467

CORRUGATOR (Supercilii)

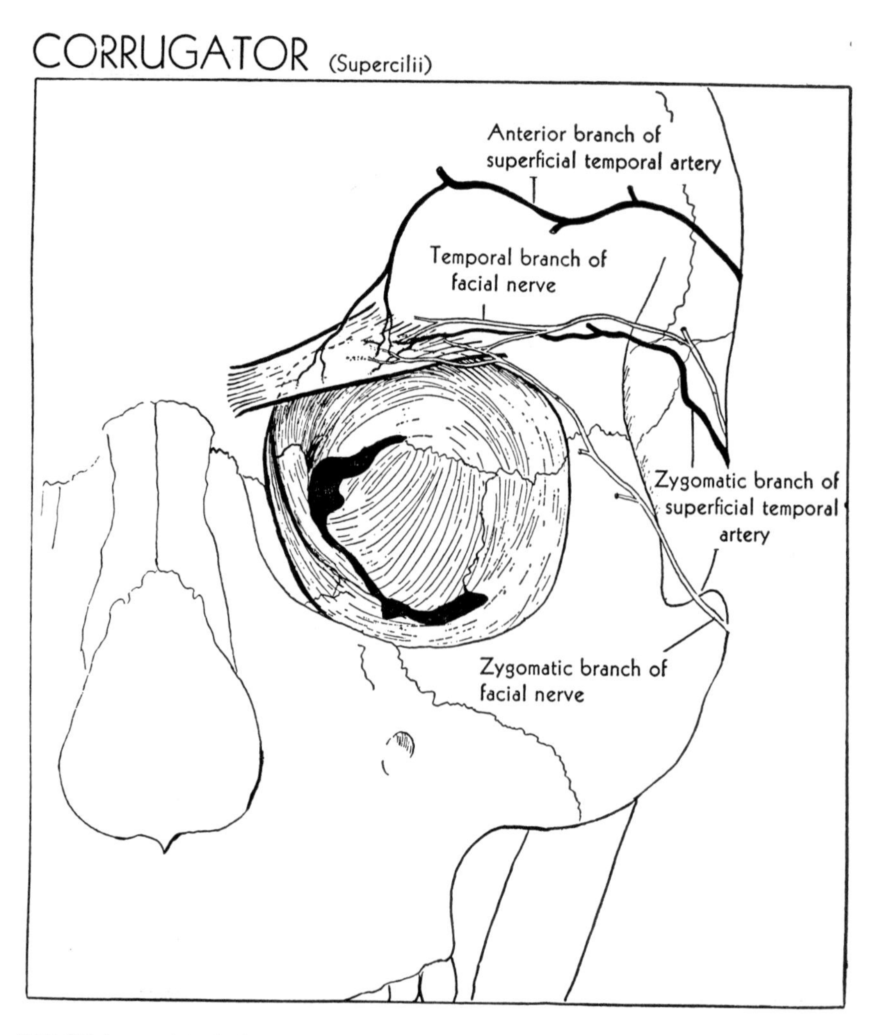

ORIGIN: Medial part of supraorbital margin
INSERTION: Skin of medial half of eyebrow
FUNCTION: Draws eyebrows downward and medially, produces wrinkles in frowning; principal muscle in expression of suffering
NERVE: Zygomatic and temporal branches of facial
ARTERY: Zygomatic and anterior branches of superficial temporal

REFERENCES:	GRAY	GRANT'S ATLAS
Muscle	382	466, 470
Nerve	382, 924, 931	467
Artery	586	Not shown

PROCERUS (Pyramidalis nasi)

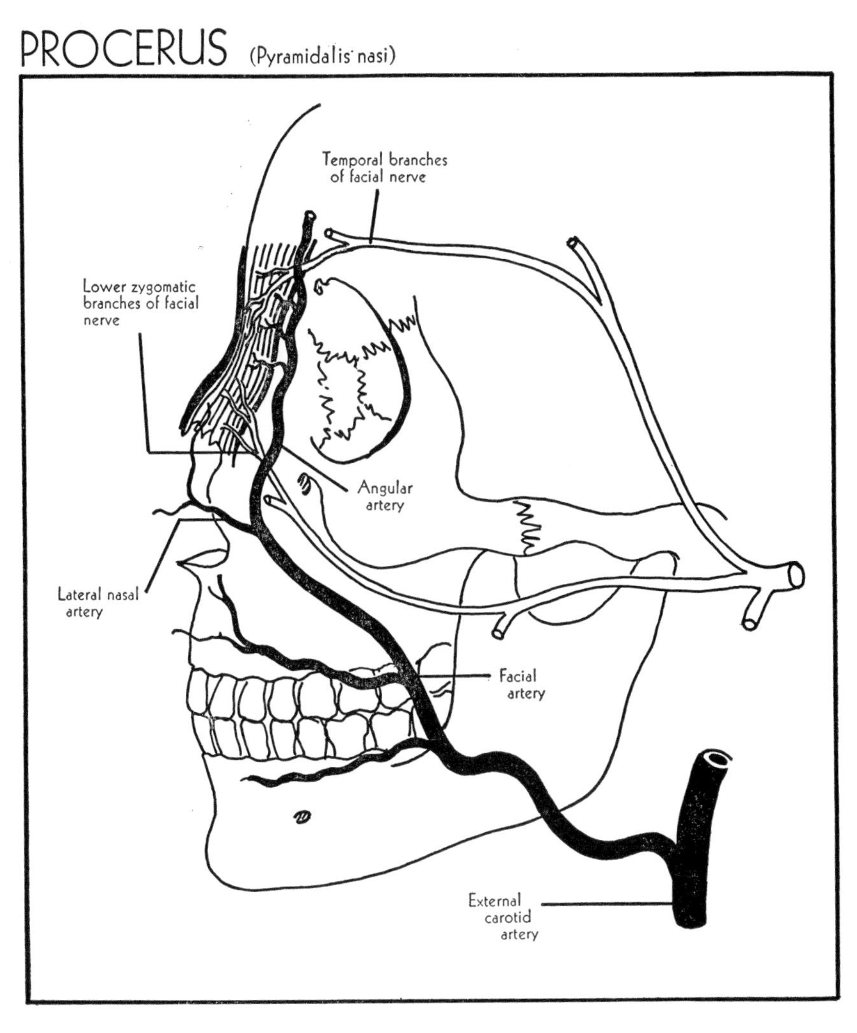

ORIGIN: Fascia covering lower parts of nasal bone and upper part of lateral nasal cartilage
INSERTION: Skin between and above eyebrow
FUNCTION: Draws down medial angle of eyebrows, produces transverse wrinkles over bridge of nose
NERVE: Temporal, lower zygomatic and buccal branches of facial
ARTERY: Angular and lateral nasal branch of facial

REFERENCES: GRAY	GRANT'S ATLAS
Muscle 382	466, 470
Nerve 383, 924, 931	467
Artery 584	466

COMPRESSOR NARIS (Nasalis, transverse part)

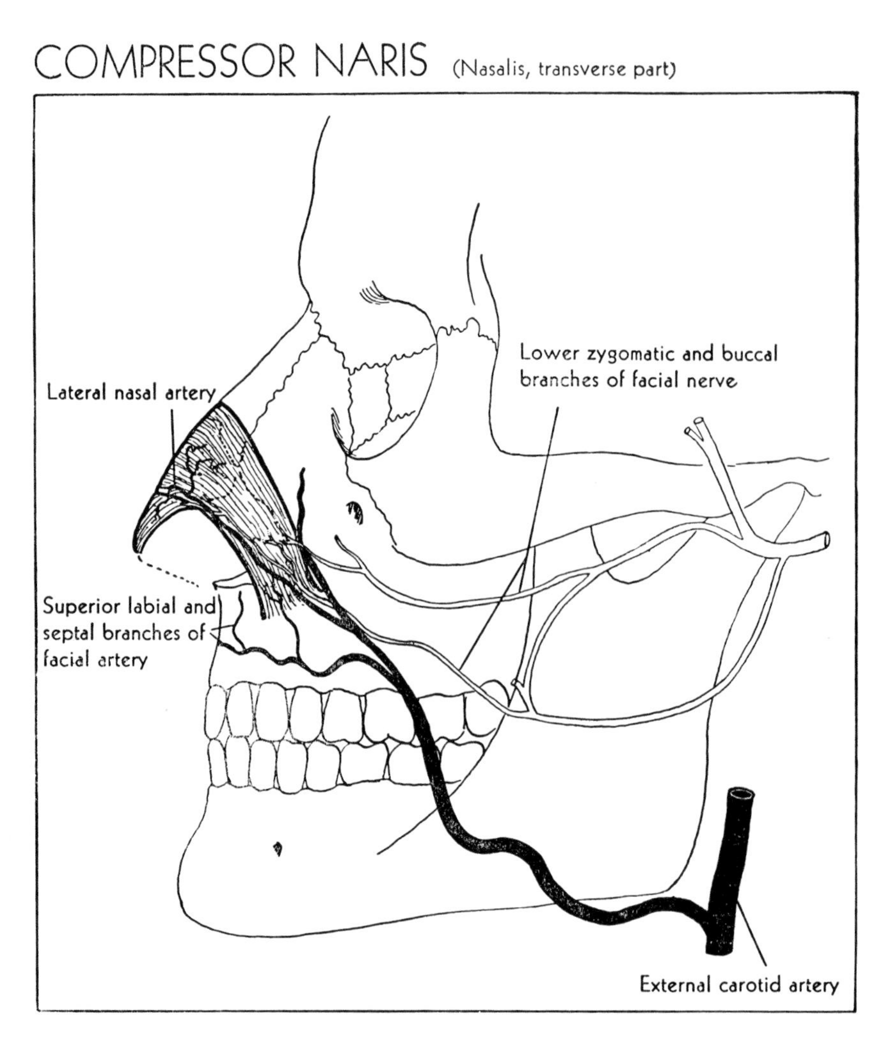

ORIGIN: Canine eminence above and lateral to incisive fossa of maxilla
INSERTION: Into aponeurosis on nasal cartilages
FUNCTION: Draws ala of nose toward septum; compressor of nostrils
NERVE: Lower zygomatic and buccal branches of facial
ARTERY: Superior labial, septal, and lateral nasal branches of facial

REFERENCES:	GRAY	GRANT'S ATLAS
Muscle	382	466
Nerve	383, 924, 931	467
Artery	584	466

DILATATOR NARIS (Nasalis, alar part)

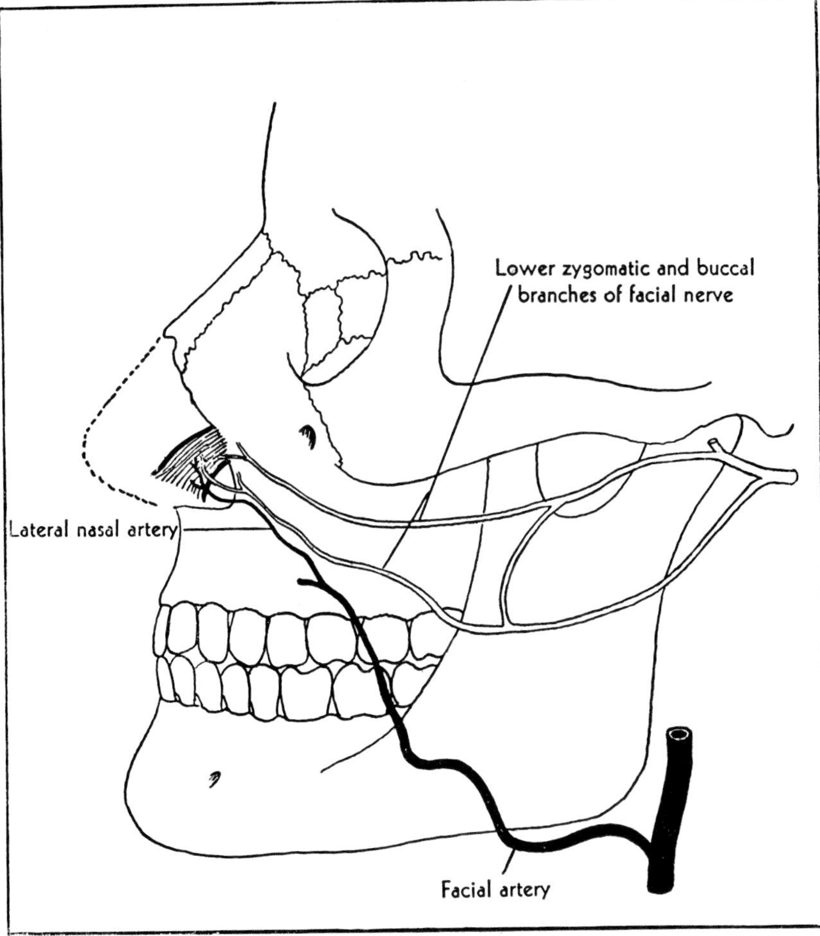

ORIGIN: Margin of nasal notch of maxilla and lesser alar cartilage
INSERTION: Skin near the margin of nostril
FUNCTION: Enlarges nasal aperture
NERVE: Zygomatic and buccal branches of facial
ARTERY: Lateral nasal branch of facial
REFERENCES: GRAY

Muscle 382
Nerve 383, 924, 931
Artery 584

GRANT'S ATLAS
Not listed in 6th edition

DEPRESSOR SEPTI

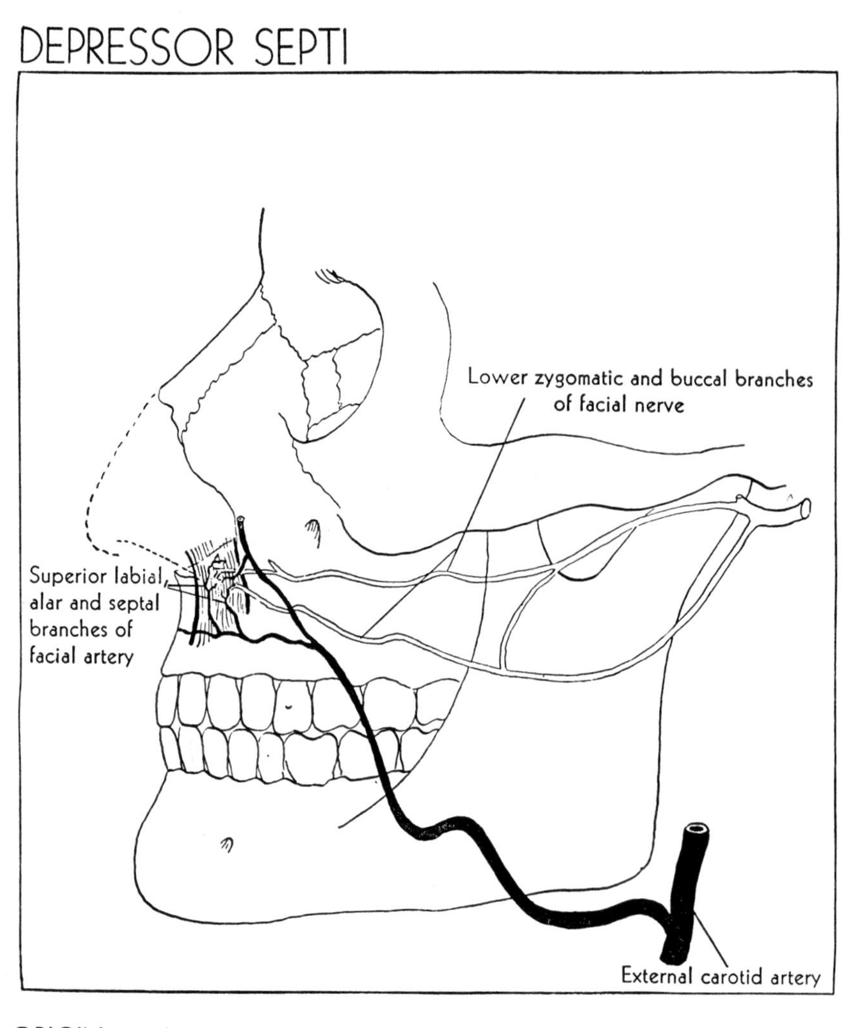

ORIGIN: Incisive fossa of maxilla
INSERTION: Septum and back part of ala of nose
FUNCTION: Narrows nostril, draws septum downward
NERVE: Lower zygomatic and buccal branches of facial
ARTERY: Superior labial septal and alar branches of facial
REFERENCES: GRAY

Muscle 382
Nerve 383, 924, 931
Artery 584

GRANT'S ATLAS
Not listed in 6th edition

LEVATOR LABII SUPERIORIS (Including alaeque nasi)

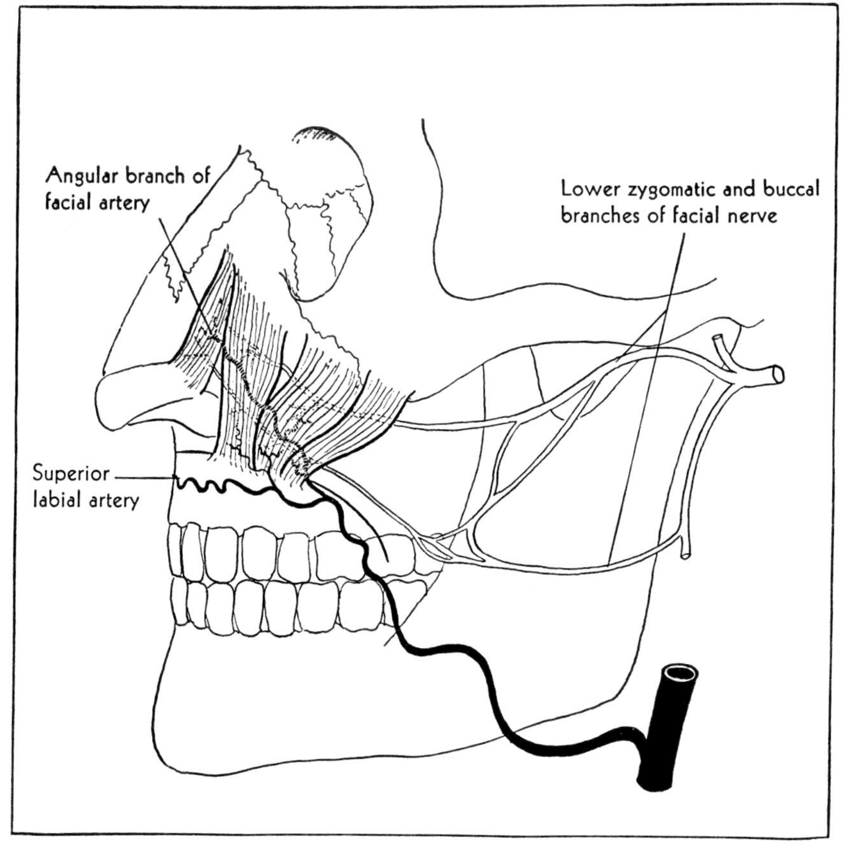

ORIGIN: Angular head, from upper part of frontal process of maxilla; infra-orbital head, from margin of orbit above infra-orbital foramen; zygomatic head, from malar surface of zygomatic bone

INSERTION: Angular head, into greater alar cartilage, skin of nose and lateral part of upper lip; infra-orbital head, into muscular substance of upper lip between angular head and caninus; zygomatic head, into skin of naso-labial groove and upper lip

FUNCTION: Angular head elevates upper lip and dilates nostril; infra-orbital head raises angle of mouth; zygomatic head elevates upper lip laterally

NERVE: Zygomatic and buccal branches of facial

ARTERY: Superior labial and angular branches of facial

REFERENCES:

	GRAY	GRANT'S ATLAS
Muscle	383	466, 470
Nerve	383, 924, 931	467
Artery	584	466

LEVATOR ANGULI ORIS (Caninus)

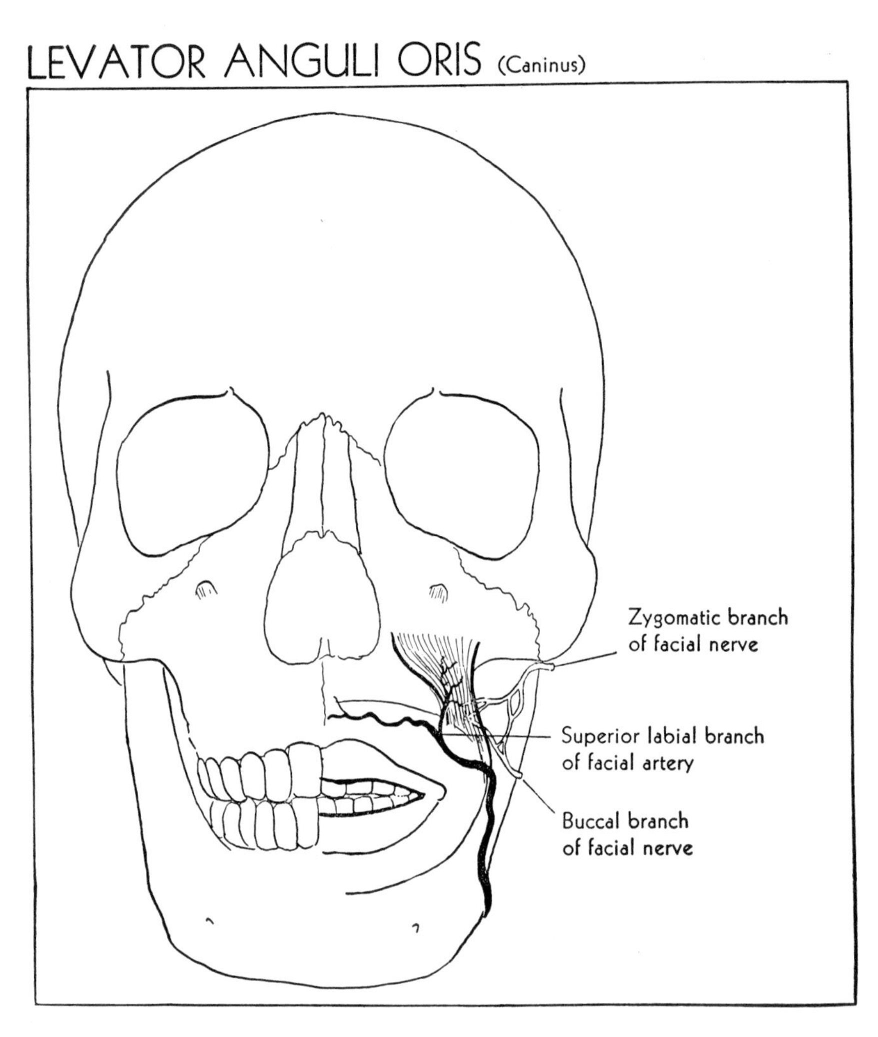

ORIGIN: Canine fossa of maxilla immediately below infra-orbital foramen and under cover of zygomatic head of quadratus labii superioris

INSERTION: Angle of mouth; fibers intermingle with orbicularis oris, depressor anguli oris and zygomaticus

FUNCTION: Elevates angle of mouth

NERVE: Zygomatic and buccal branches of facial

ARTERY: Superior labial branch of facial

REFERENCES:	GRAY	GRANT'S ATLAS
Muscle	383	470
Nerve	383, 924, 931	467
Artery	584	466

ZYGOMATICUS (major)

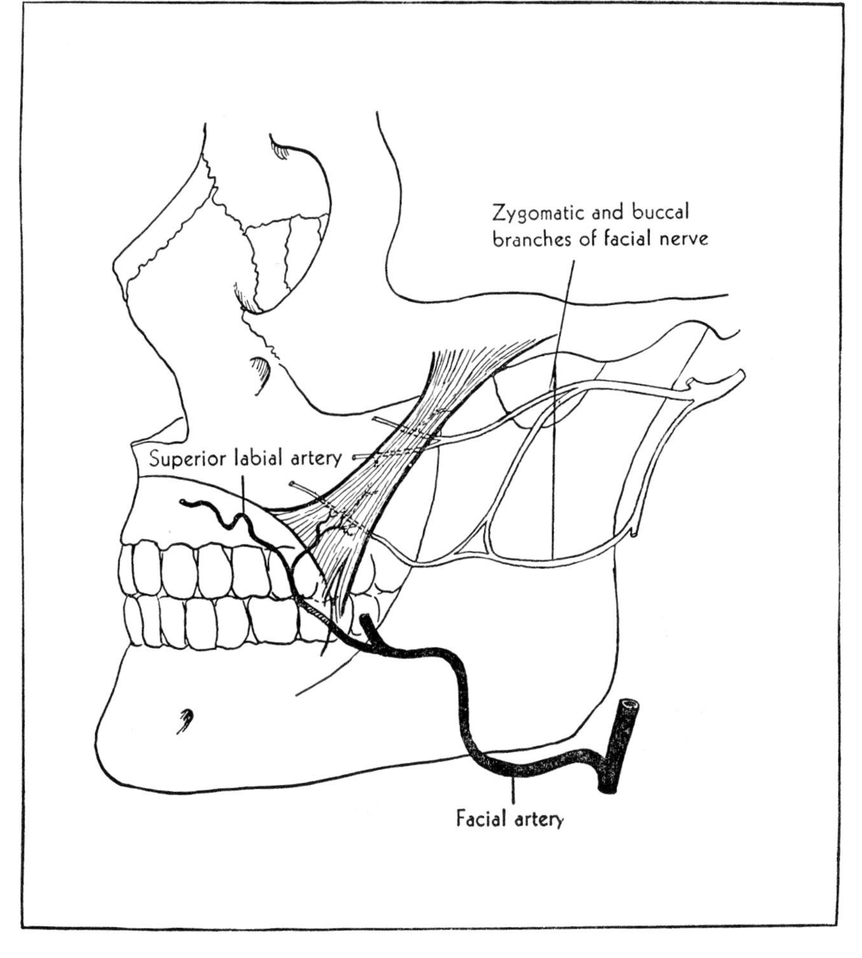

ORIGIN: From zygomatic portion of zygomatic arch
INSERTION: Into angle of mouth, mingling with orbiculars oris, caninus and depressor anguli oris
FUNCTION: Draws angle of mouth backward and upward (laughing)
NERVE: Zygomatic and buccal branches facial
ARTERY: Superior labial branch of facial

REFERENCES: GRAY	GRANT'S ATLAS
Muscle 383	466, 467, 470
Nerve 383, 924, 931	467
Artery 584	466, 467

RISORIUS

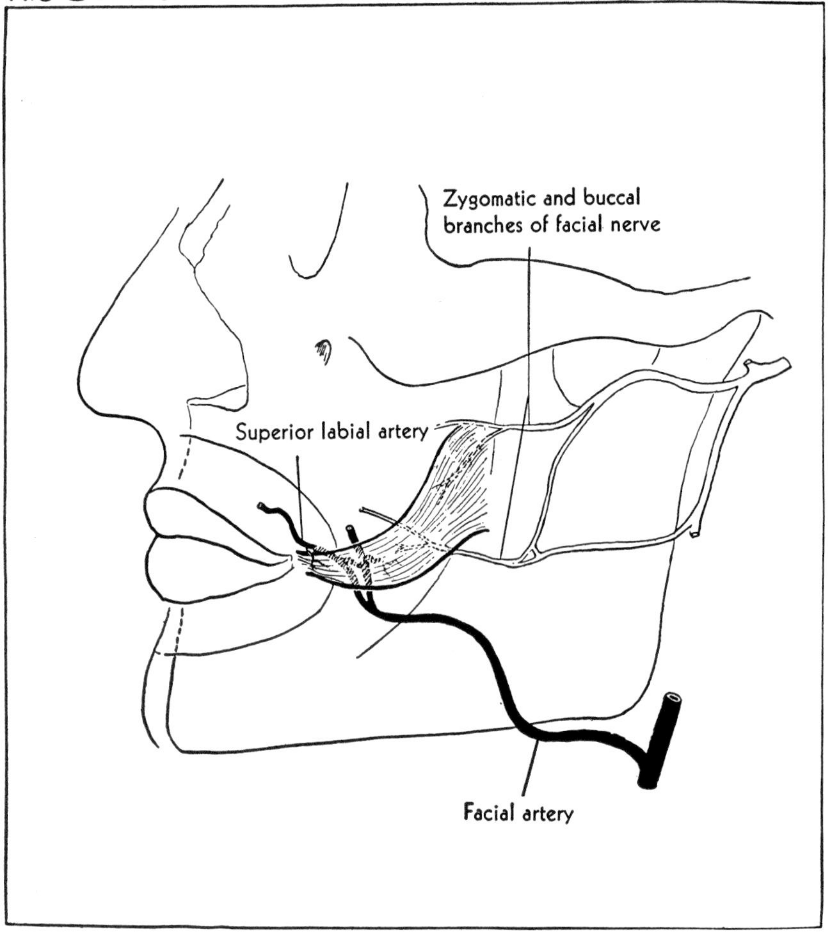

ORIGIN: Fascia over masseter superficial to platysma
INSERTION: Skin at angle of mouth
FUNCTION: **Retracts angle of mouth; (grinning)**
NERVE: Zygomatic and buccal branches of facial
ARTERY: Superior labial branch of facial

REFERENCES: GRAY
- Muscle 383
- Nerve 384, 924, 931
- Artery 584

GRANT'S ATLAS
Not listed in 6th edition

DEPRESSOR LABII INFERIORIS (Quadratus labii inferioris)

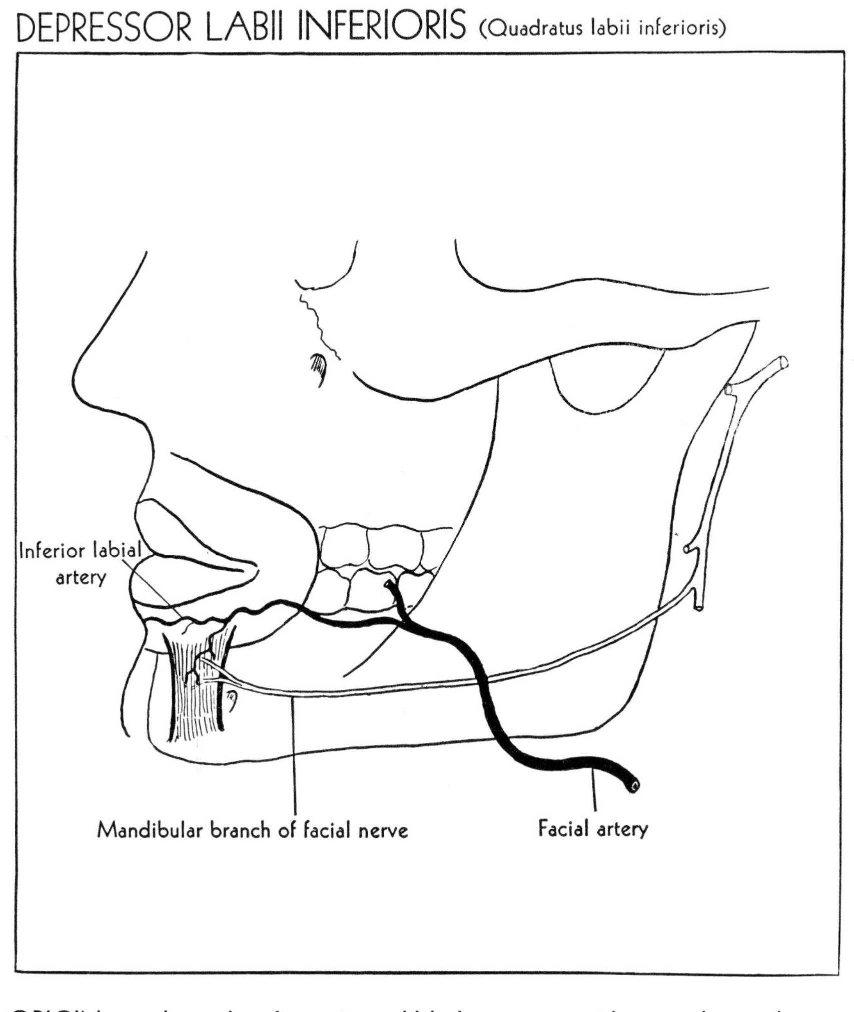

ORIGIN: Lateral surface of mandible between symphysis and mental foramen deep to anguli oris

INSERTION: Skin of lower lip, mingling with orbicularis oris, medial fibers joining those of opposite side

FUNCTION: Depresses lower lip and draws it lateralward (irony)

NERVE: Mandibular and buccal branches of facial

ARTERY: Inferior labial branch of facial

REFERENCES:	GRAY	GRANT'S ATLAS
Muscle	383	466, 526
Nerve	384, 924, 931	467
Artery	584	466, 467

DEPRESSOR ANGULI ORIS (Triangularis)

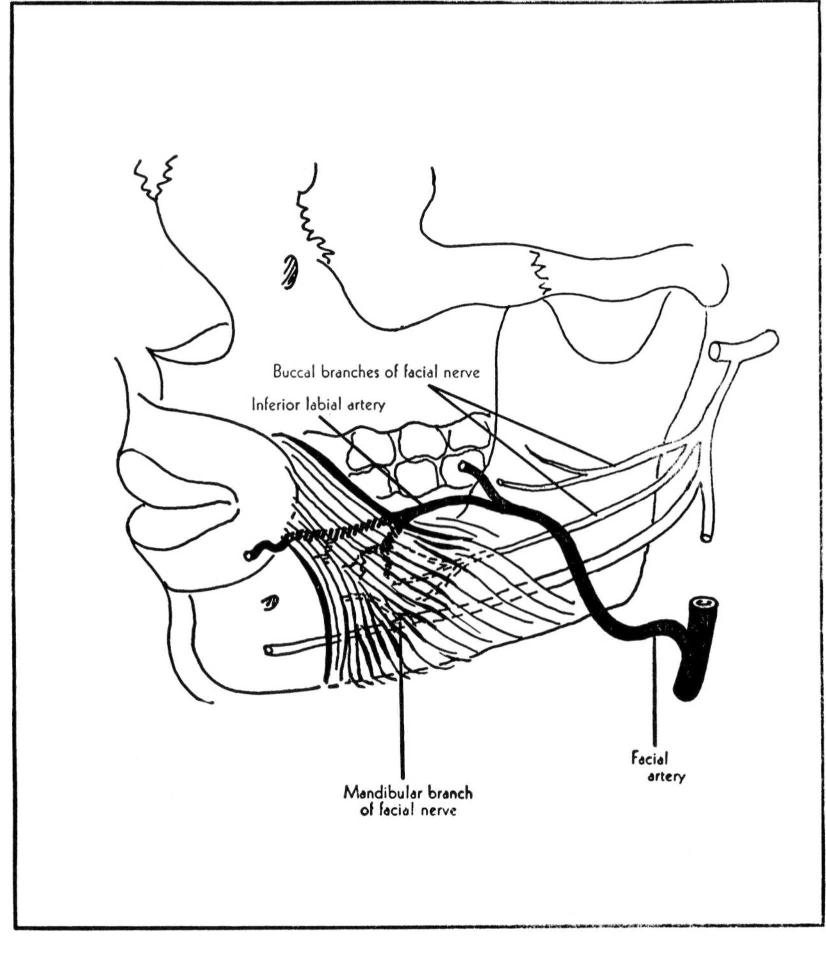

ORIGIN: Continuous with platysma on oblique line of mandible
INSERTION: Angle of mouth into orbicularis and skin
FUNCTION: Depresses angle of mouth; associated with grief
NERVE: Mandibular and buccal branches of facial
ARTERY: Inferior labial branch of facial

REFERENCES:	GRAY	GRANT'S ATLAS
Muscle	384	466, 467, 470, 526
Nerve	384, 924, 931	467
Artery	584	466, 467

MENTALIS

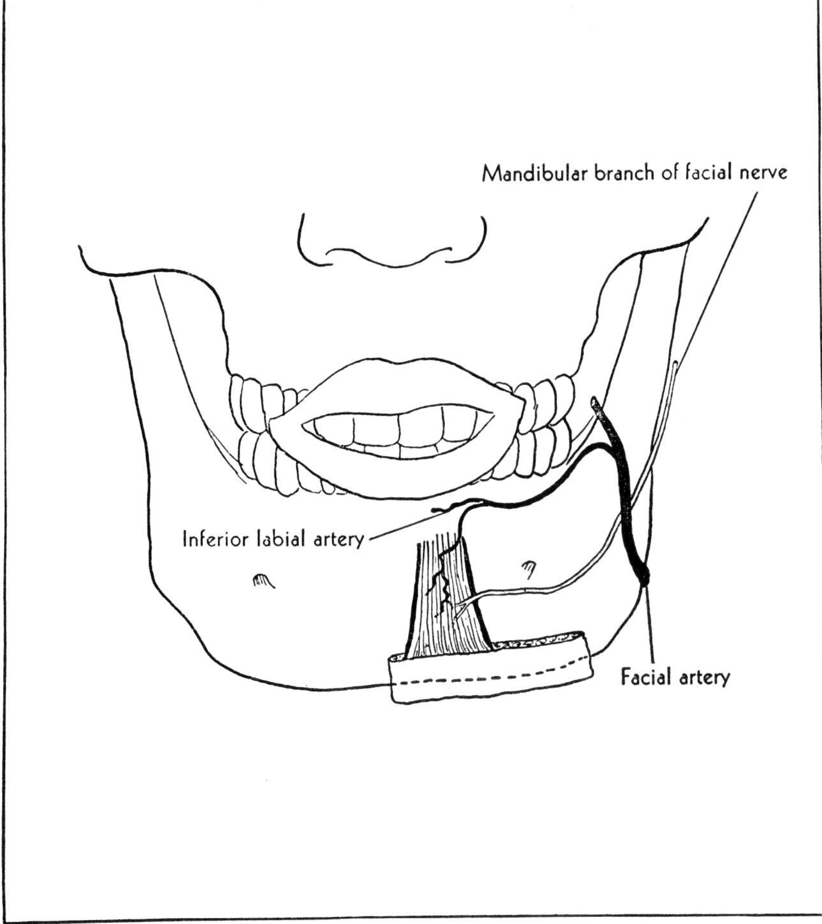

ORIGIN: Incisive fossa of mandible
INSERTION: Skin of chin
FUNCTION: Raises and protrudes lower lip, wrinkles skin, expresses doubt or disdain
NERVE: Mandibular branch of facial
ARTERY: Inferior labial branch of facial

REFERENCES: GRAY	GRANT'S ATLAS
Muscle 384	466, 470
Nerve 384, 924, 931	467
Artery 584	Not shown

ORBICULARIS ORIS

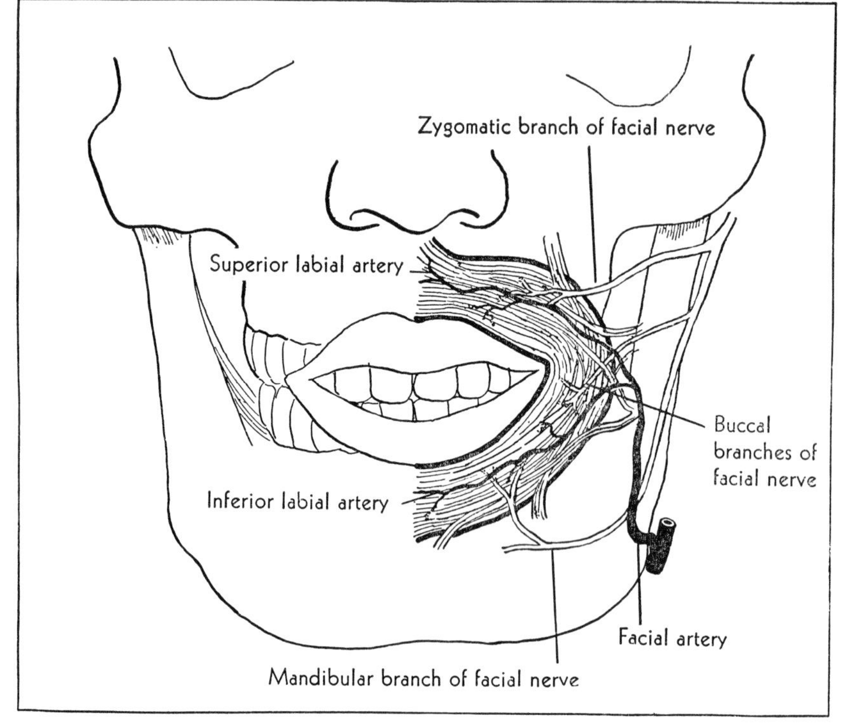

ORIGIN and INSERTION: Sphincter muscle formed by various facial muscles converging on the mouth, and by its own proper fibers. Deep stratum derived from buccinator decussates at angle of mouth, fibers from maxilla pass to lower lip, those from mandible to upper lip. Upper and lower buccinator fibers pass across lips without decussation. More superficially levator anguli oris and depressor anguli oris fibers cross at angle of mouth, those from levator to lower lip, those from depressor to upper. Upper fibers insert near median line. Fibers from levator labii superioris, zygomaticus, depressor labii inferioris intermingle with transverse fibers. Proper fibers of lips run obliquely under skin surface to mucous membrane. A lateral band of proper fibers arises from alveolar border of maxilla opposite lateral incisors, arches laterally to continue with fibers at mouth angle. A medial band connects upper lip with nasal septum

FUNCTION: Compression, contraction and protrusion of lips; facial expression

NERVE: Lower zygomatic, buccal and mandibular branches of facial

ARTERY: Inferior and superior labial branches of facial

REFERENCES:

	GRAY	GRANT'S ATLAS
Muscle	**384**	**470**
Nerve	**385, 924, 931**	**467**
Artery	**584**	**Not shown**

BUCCINATOR

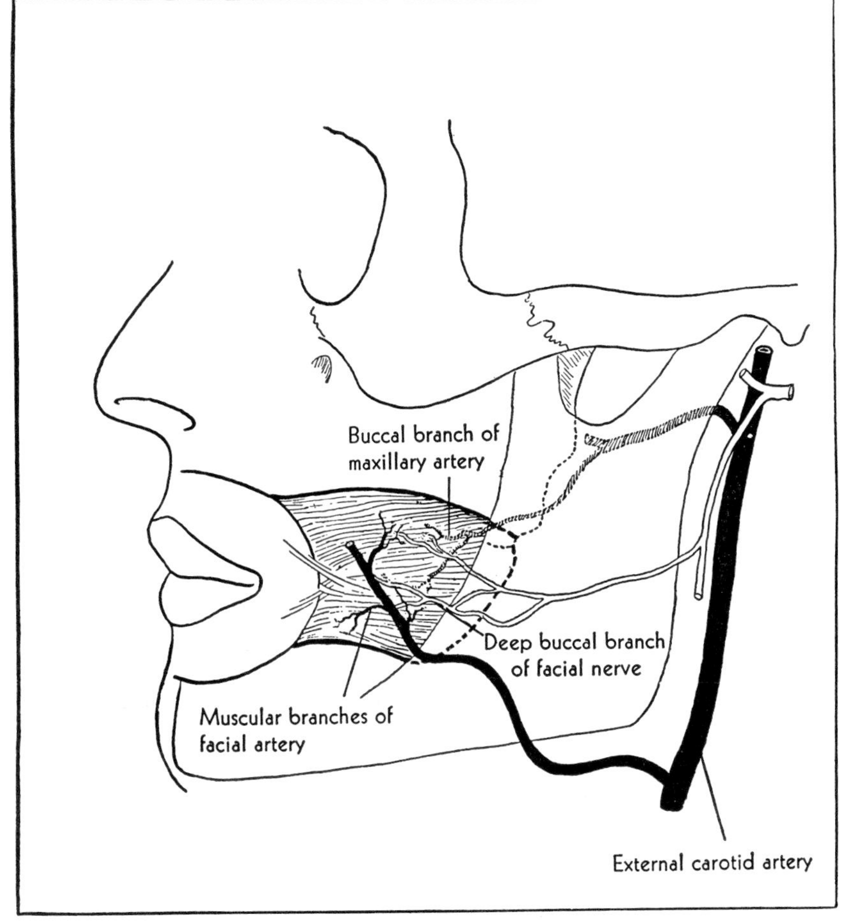

ORIGIN: Alveolar process of maxilla and mandible opposite sockets of molar teeth; anterior border of pterygo-mandibular raphé

INSERTION: Fibers converge toward angle of mouth, where they blend with orbicularis oris; upper fibers pass to lower segment of orbicularis, lower pass to upper segment of that muscle

FUNCTION: Compresses cheeks, expels air between lips, aids in mastication

NERVE: Deep buccal branches of facial

ARTERY: Muscular branches of facial, buccal branch of maxillary

REFERENCES:	GRAY	GRANT'S ATLAS
Muscle	385	465, 559, 574, 591
Nerve	386, 924, 931	467
Artery	584, 590	467, 559

TEMPORALIS

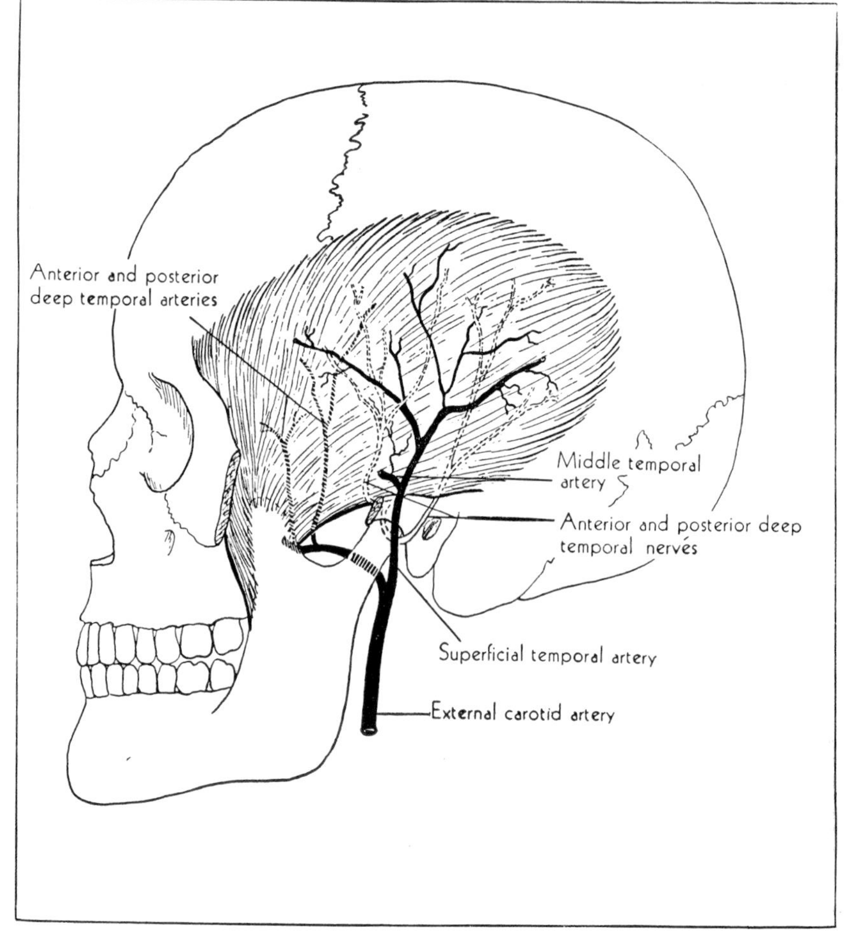

ORIGIN: Floor of temporal fossa and temporal fascia

INSERTION: Anterior border of coronoid process and anterior border of ramus of mandible

FUNCTION: Elevates jaw, retracts mandible, clenches teeth

NERVE: Deep temporal branches of anterior trunk of mandibular division of trigeminal

ARTERY: Middle temporal branches of superficial temporal; anterior and posterior deep temporal branch of maxillary

REFERENCES:

	GRAY	GRANT'S ATLAS
Muscle	387	551, 559, 579, 591
Nerve	389, 919, 920	559, 560
Artery	586, 590	466, 467, 559

MASSETER

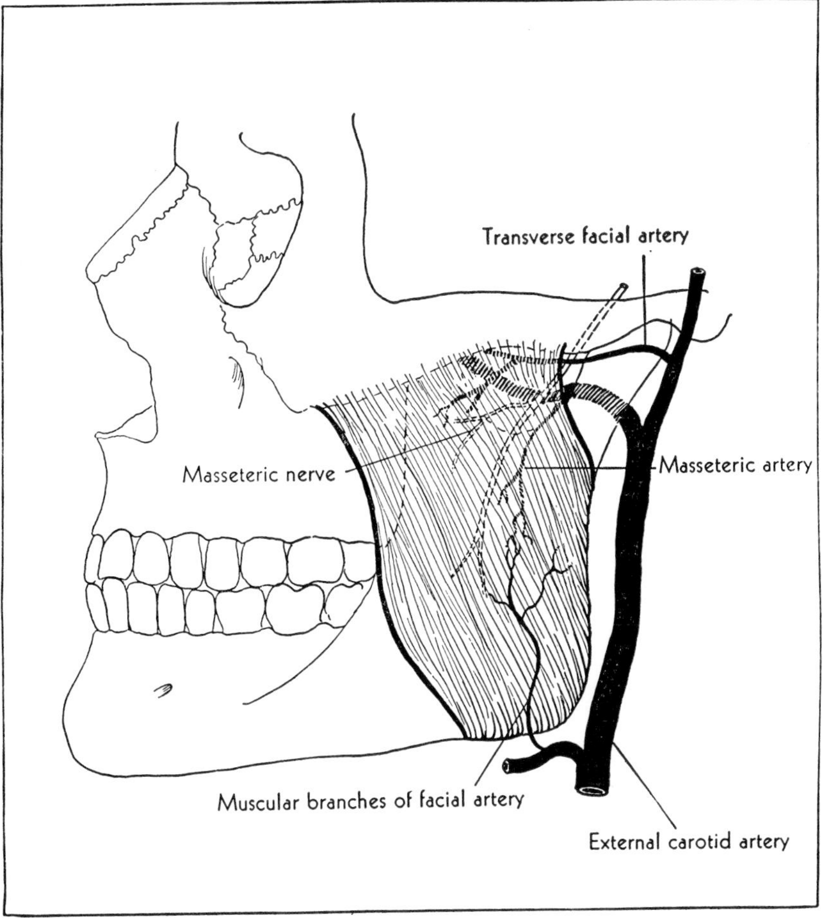

ORIGIN: Superficial portion from anterior two-thirds of lower border of zygomatic arch; deep portion from medial surface of zygomatic arch

INSERTION: Lateral surface of coronoid process of mandible, upper half of ramus and angle of mandible

FUNCTION: Elevates jaw, clenches teeth

NERVE: Masseteric nerve from anterior trunk of mandibular division of trigeminal

ARTERY: Transverse facial branch of superficial temporal; masseteric branch of maxillary; muscular branches of facial

REFERENCES:	GRAY	GRANT'S ATLAS
Muscle	389	465-467, 579, 591
Nerve	389, 919, 920	551, 559, 560
Artery	584, 586, 590	466, 467

PTERYGOIDEUS MEDIALIS (Internus)

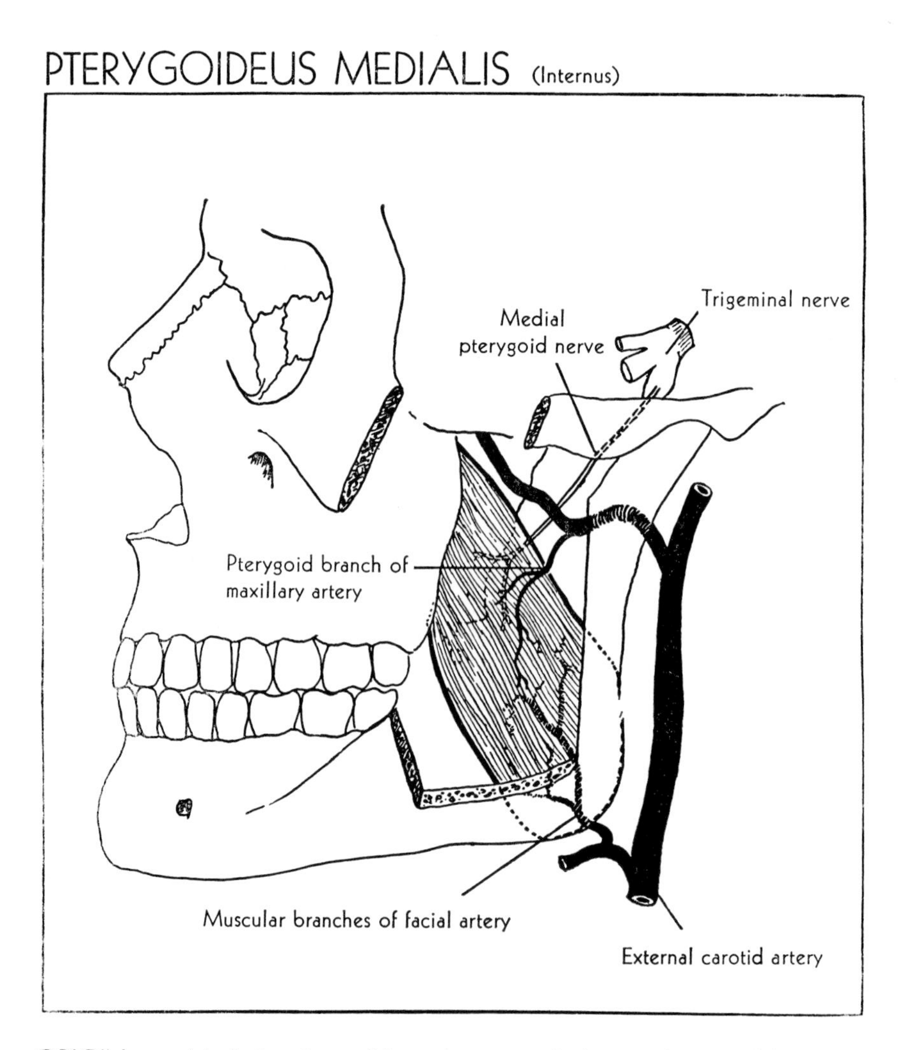

ORIGIN: Medial surface of lateral pterygoid plate and pyramidal process of palatine bone; small slip from tuberosity of maxilla

INSERTION: Lower and back part of medial surface of ramus and angle of mandible

FUNCTION: Protracts and elevates lower jaw; assists in rotary motion while chewing

NERVE: Medial pterygoid from mandibular division of trigeminal

ARTERY: Muscular branches of facial; pterygoid branches of maxillary

REFERENCES:

	GRAY	GRANT'S ATLAS
Muscle	389	559, 584, 598
Nerve	389, 919, 920	Not shown
Artery	584, 590	Not shown

PTERYGOIDEUS LATERALIS (Externus)

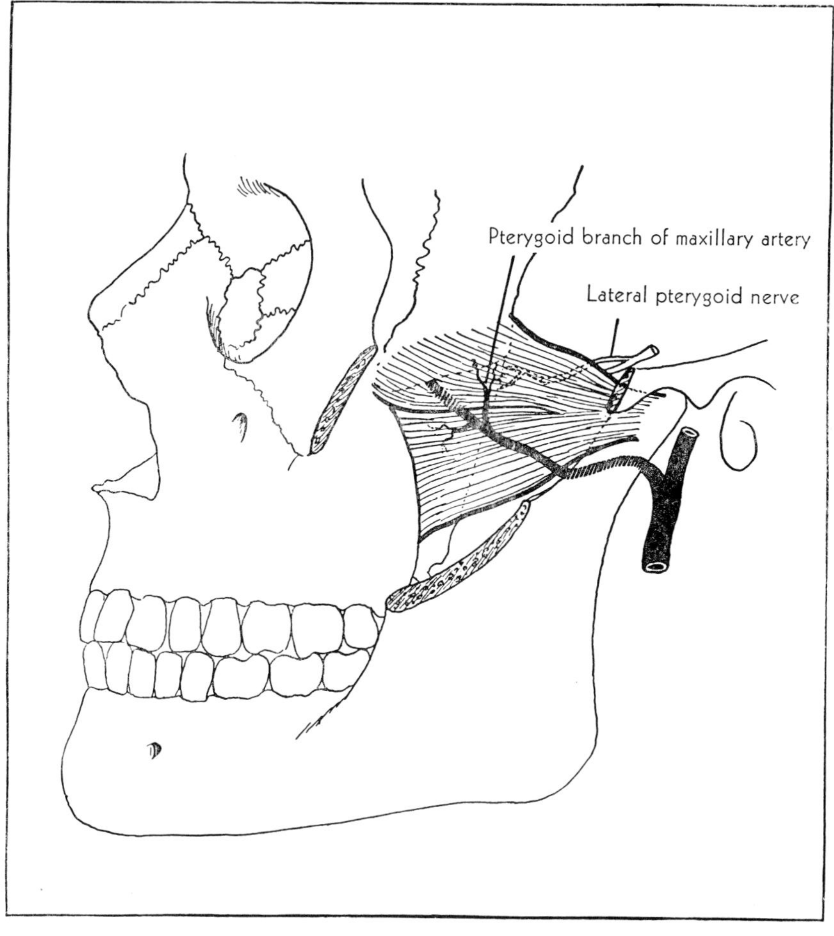

ORIGIN: Upper head from infratemporal surface of greater wing of sphenoid; lower head from lateral surface of lateral pterygoid plate

INSERTION: Front of neck of mandibular condyle and capsule of mandibular joint

FUNCTION: Protrudes mandible, pulls articular disc forward; assists in rotary motion while chewing

NERVE: Lateral pterygoid, branch of anterior division of mandibular

ARTERY: Pterygoid branch of maxillary

REFERENCES:

	GRAY	GRANT'S ATLAS
Muscle	389	559, 579
Nerve	389, 919, 920	560
Artery	590	559

PLATYSMA

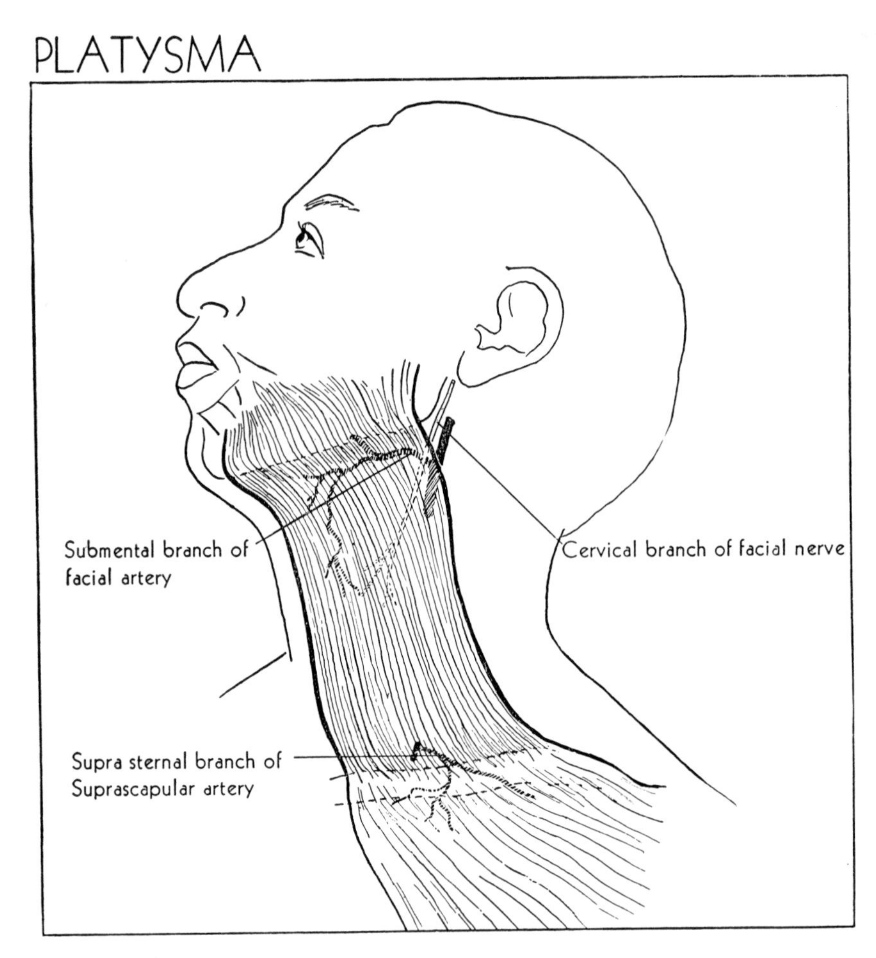

ORIGIN: Upper pectoral and deltoid regions by bundles from superficial fascia. These fibers cross the clavicle and pass obliquely upward and medially along the sides of the neck

INSERTION: Anterior fibers of one side interlace below the chin with those of the other and connect with depressor labii inferioris and depressor anguli oris. Posterior fibers pass across the angle of the jaw and some insert into the mandible, others pass to skin of lower part of face, many blending with muscles at angle and lower part of mouth

FUNCTION: Depresses lower jaw and lip, tenses and ridges skin of neck

NERVE: Cervical branch of facial

ARTERY: Submental branch of facial, supra-sternal branch of suprascapular (off thyrocervical trunk)

REFERENCES:	GRAY	GRANT'S ATLAS
Muscle	390	13, 467, 526
Nerve	390, 924, 931	467
Artery	583, 605-606	467

STERNOCLEIDOMASTOIDEUS

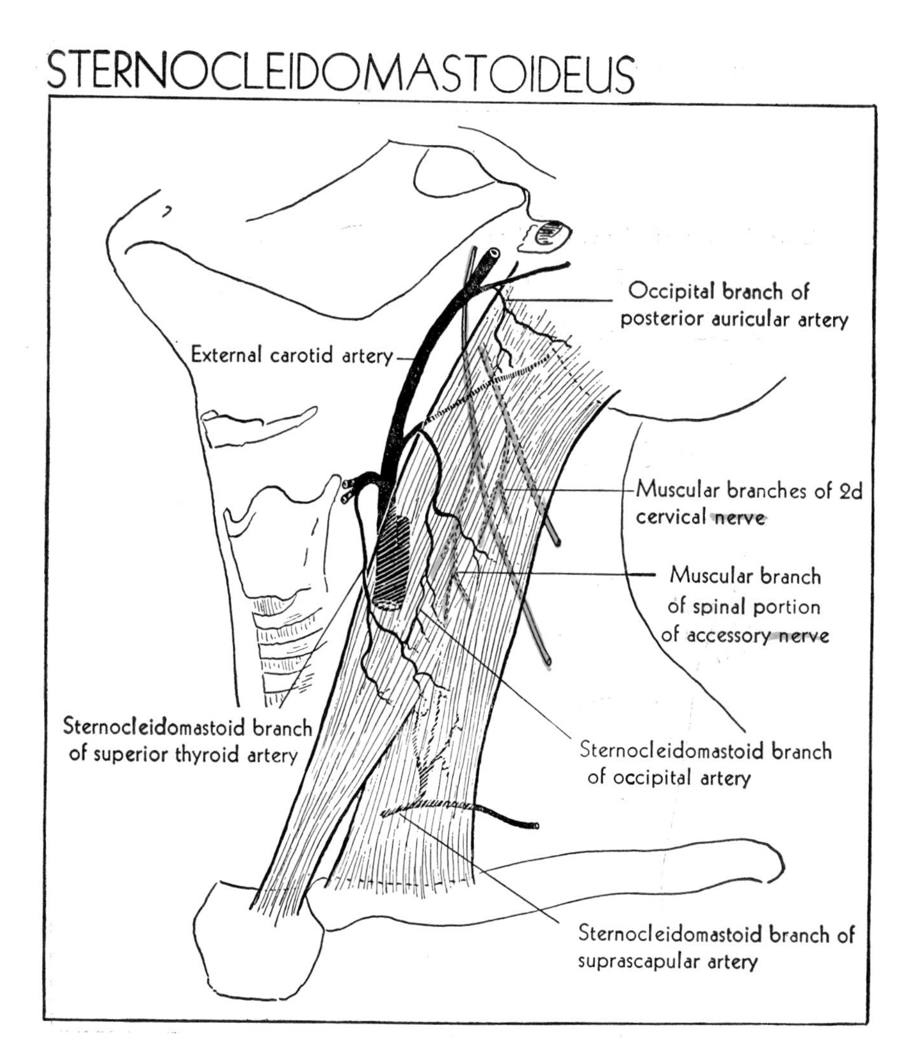

ORIGIN: Sternal head, anterior surface of manubrium; clavicular head, upper surface of medial 3d of clavicle.

INSERTION: Lateral surface of mastoid process; lateral half of superior nuchal line of occipital bone

FUNCTION: Singly, draws head toward shoulder and rotates it pointing chin cranially and to opposite side; together, flex head; raise thorax when head is fixed

NERVE: 2d cervical and spinal portion of accessory

ARTERY: Sternocleidomastoid branch of superior thyroid and occipital, muscular of suprascapular, occipital of posterior auricular

REFERENCES:

	GRAY	GRANT'S ATLAS
Muscle	395	406, 473, 542
Nerve	396, 945, 955-956, 960	542, 661
Artery	581, 585, 586, 605	467, 541, 542, 552

DIGASTRICUS

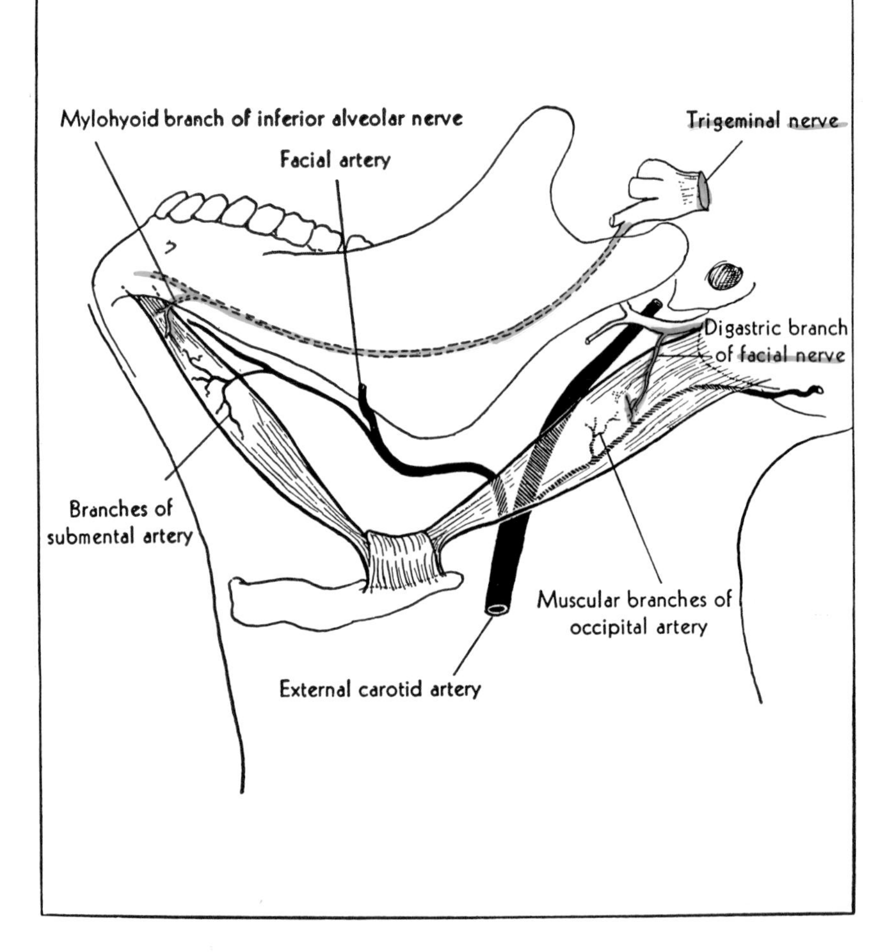

ORIGIN: Posterior belly from mastoid notch of temporal bone; anterior belly from digastric fossa of mandible

INSERTION: Both bellies by an intermediate tendon which passes through insertion of stylohyoid muscle and is attached to side of body and greater cornu of hyoid bone by a fibrous loop

FUNCTION: Raises hyoid bone and base of tongue, steadies hyoid bone

NERVE: Posterior belly, branches from facial; anterior belly, mylo-hyoid branch of inferior alveolar

ARTERY: Posterior belly, muscular branches of posterior auricular, muscular branches of occipital; anterior belly, branches of submental

REFERENCES:	GRAY	GRANT'S ATLAS
Muscle	396	527, 540.2, 546, 551
Nerve	397, 923, 930	542, 546, 552, 598
Artery	583, 585	527, 542, 546, 550

STYLOHYOIDEUS

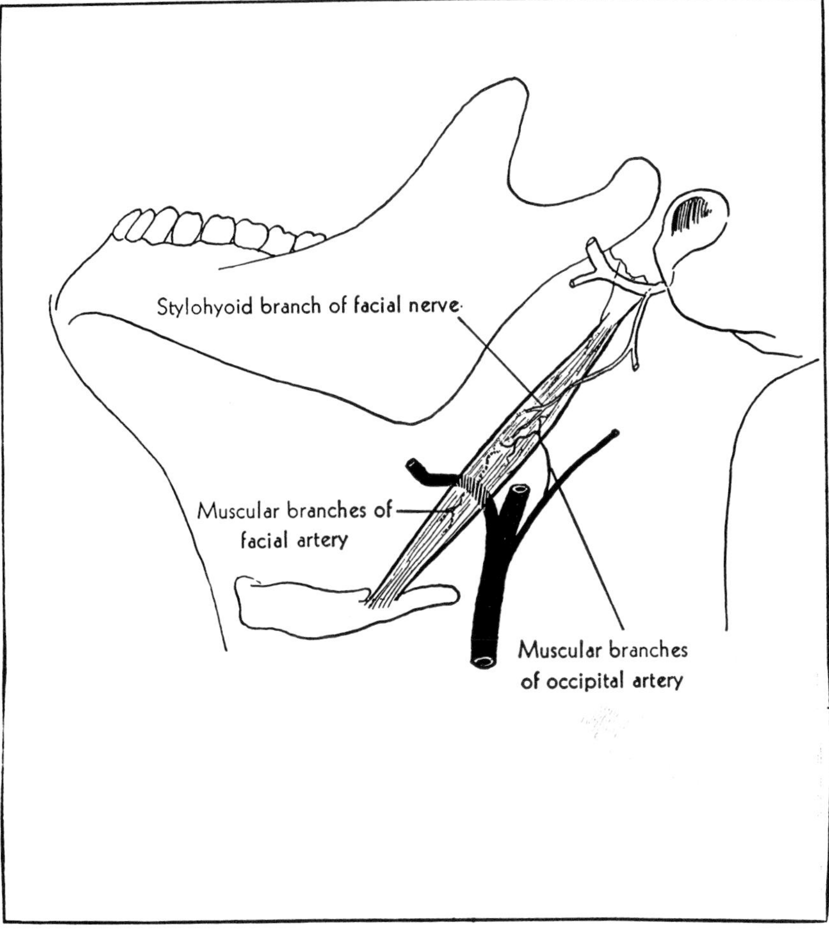

ORIGIN: Posterior border of styloid process of temporal bone near its base

INSERTION: Body of hyoid bone at junction with greater horn, just above omohyoid

FUNCTION: Elevates hyoid bone and base of tongue

NERVE: Stylo-hyoid branch from posterior trunk of facial

ARTERY: Muscular branches of facial, muscular branches of occipital

REFERENCES:	GRAY	GRANT'S ATLAS
Muscle	397	542, 551, 574
Nerve	397, 930	552
Artery	584, 585	542, 546

MYLOHYOIDEUS

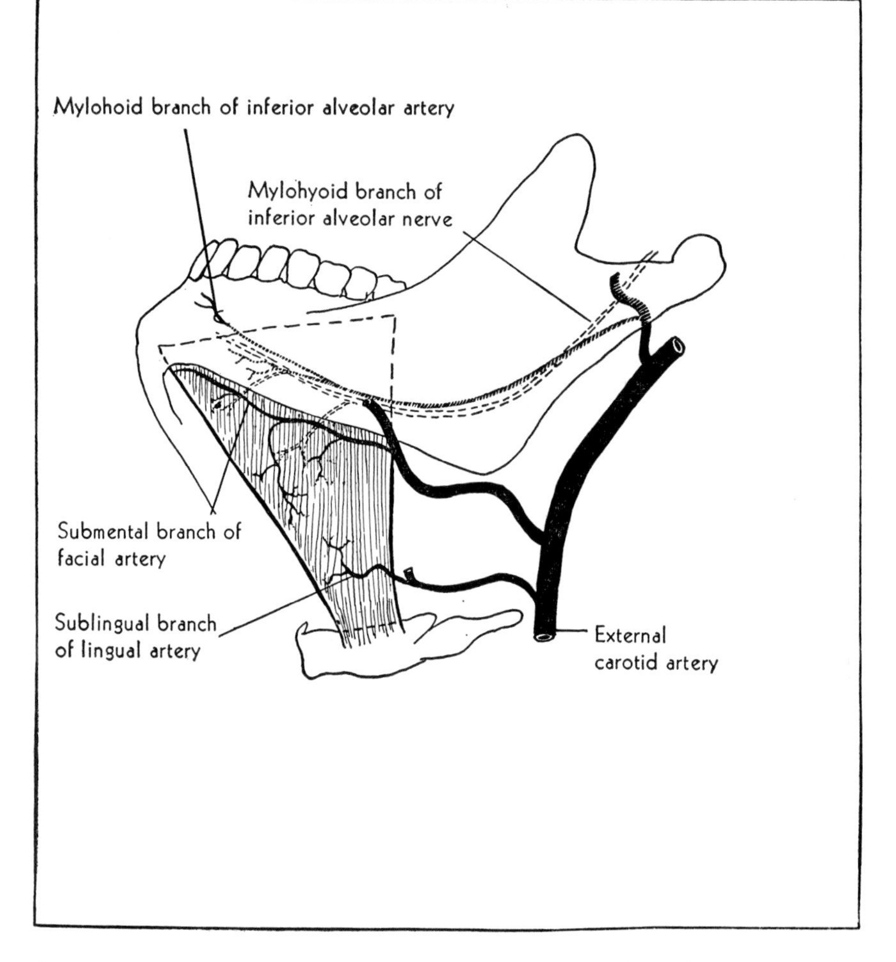

ORIGIN: Line extending from symphysis of mandible to last molar (mylohyoid line)

INSERTION: Median raphé from chin to hyoid bone and into hyoid bone

FUNCTION: Elevates hyoid bone and base of tongue, raises floor of mouth; when hyoid bone is fixed, depresses mandible

NERVE: Mylo-hyoid branch of inferior alveolar (branch of trigeminal)

ARTERY: Sublingual branch of lingual, submental branch of facial, mylo-hyoid branch of inferior alveolar

REFERENCES:	GRAY	GRANT'S ATLAS
Muscle	397	544, 547, 590, 591
Nerve	397, 923	542, 546, 598
Artery	582, 583, 589	545

GENIOHYOIDEUS

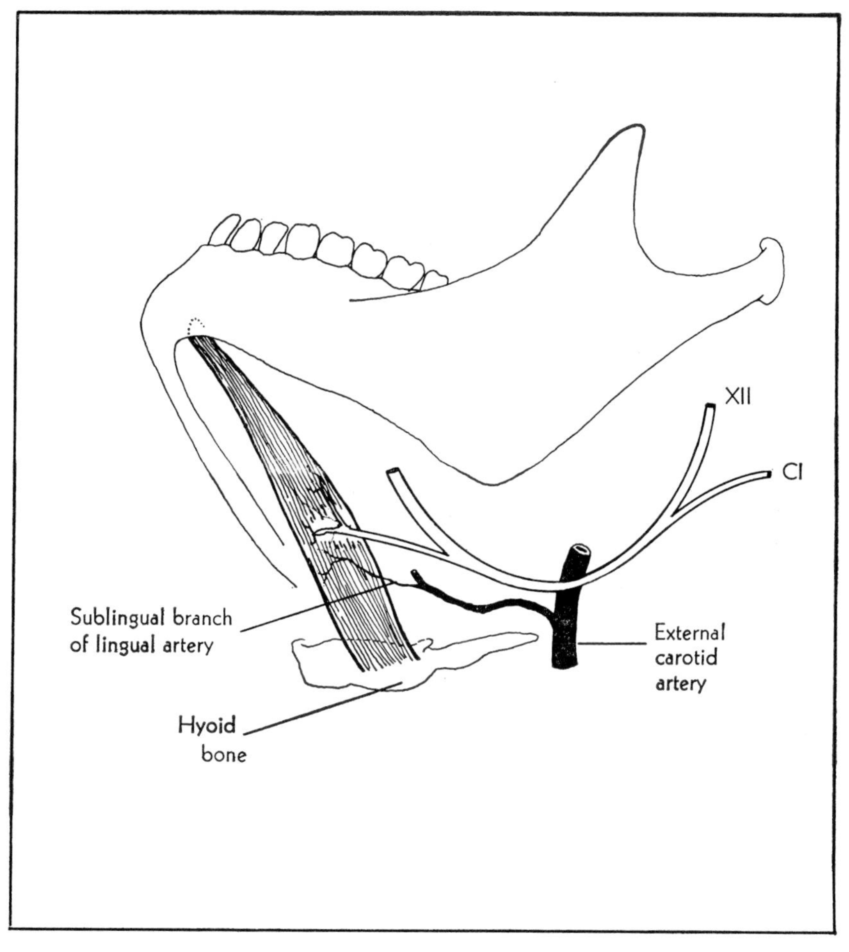

ORIGIN: Inferior genial tubercle on back of symphysis of mandible
INSERTION: Anterior surface of body of hyoid bone
FUNCTION: Elevates hyoid bone and tongue
NERVE: Branch of CI through hypoglossal
ARTERY: Sublingual branch of lingual

REFERENCES: GRAY	GRANT'S ATLAS
Muscle 397	545, 591, 595, 599
Nerve 397, 946, 955, 958	662
Artery 582	548, 549

STERNOHYOIDEUS

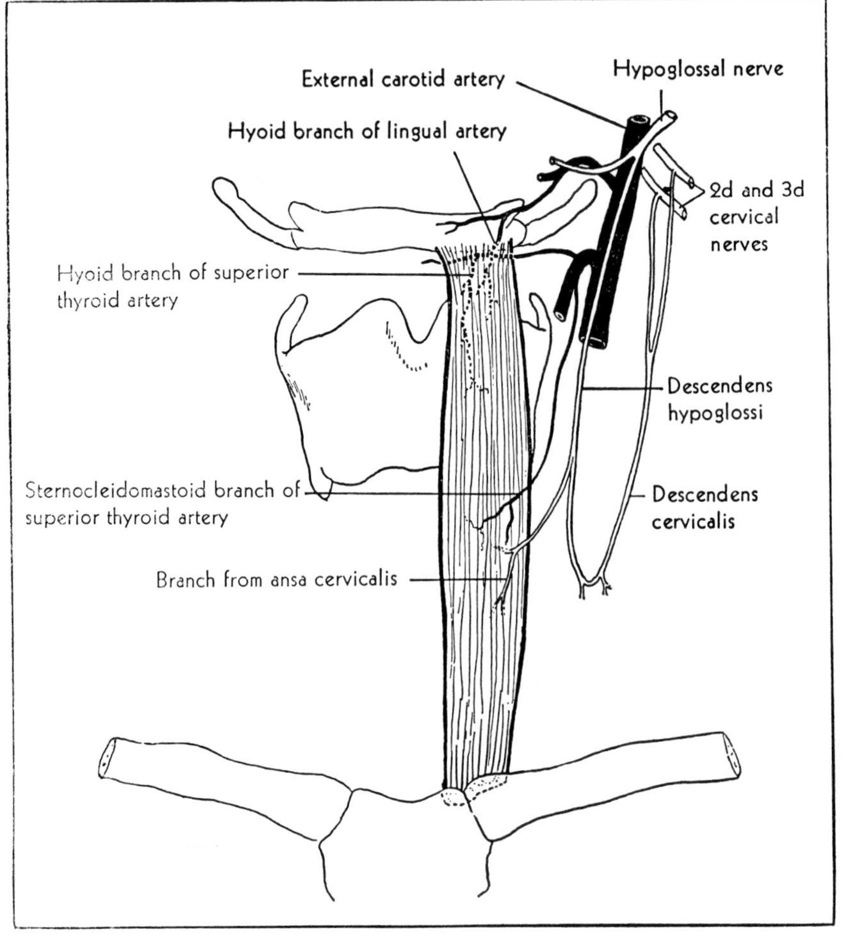

ORIGIN: Posterior surface of manubrium sterni, posterior sterno-clavicular ligament, medial end of clavicle
INSERTION: Medial part of lower border of body of hyoid bone
FUNCTION: Depresses larynx and hyoid bone, steadies hyoid bone
NERVE: Ansa cervicalis
ARTERY: Sternocleidomastoid and hyoid branches of superior thyroid, hyoid branch of lingual

REFERENCES: GRAY	GRANT'S ATLAS
Muscle 398	407, 528, 542
Nerve 398, 946, 955, 959	542, 662
Artery 579, 582	548, 549

STERNOTHYREOIDEUS

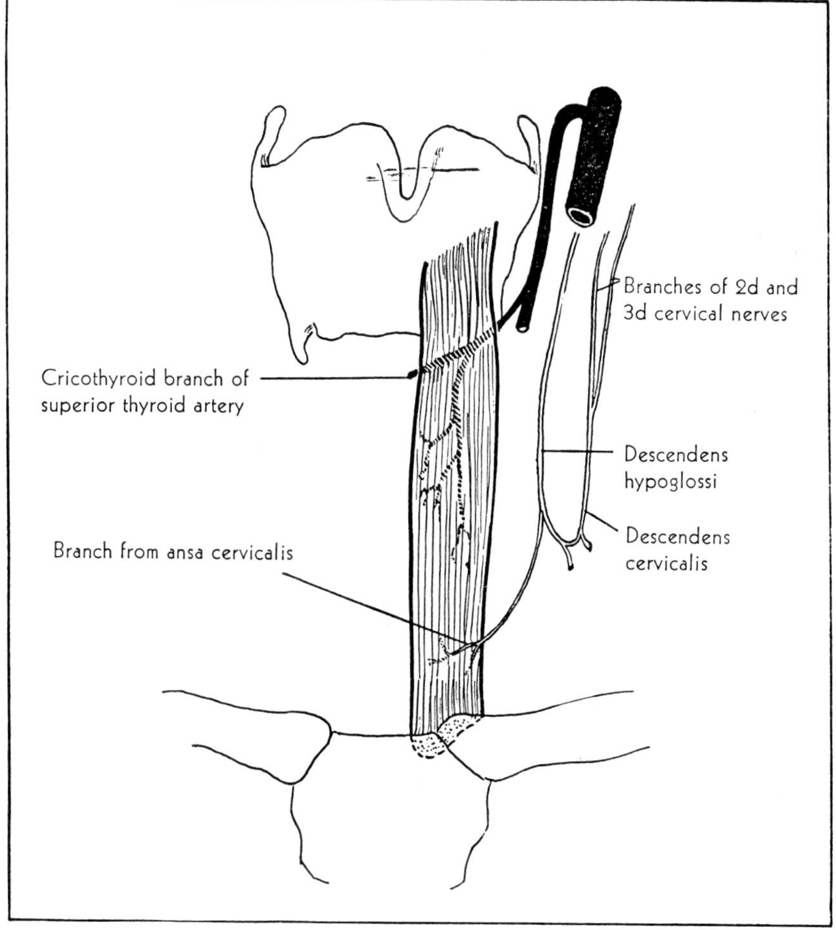

ORIGIN: Posterior surface of manubrium sterni below and deep to origin of sterno-hyoid, edge of first costal cartilage
INSERTION: Oblique line on lamina of thyroid cartilage
FUNCTION: Depresses larynx and thyroid cartilage
NERVE: Ansa cervicalis
ARTERY: Crico-thyroid branch of superior thyroid

REFERENCES:	GRAY	GRANT'S ATLAS
Muscle	398	407, 528, 529, 542
Nerve	399, 946, 955, 959	542, 662
Artery	581	529

THYREOHYOIDEUS

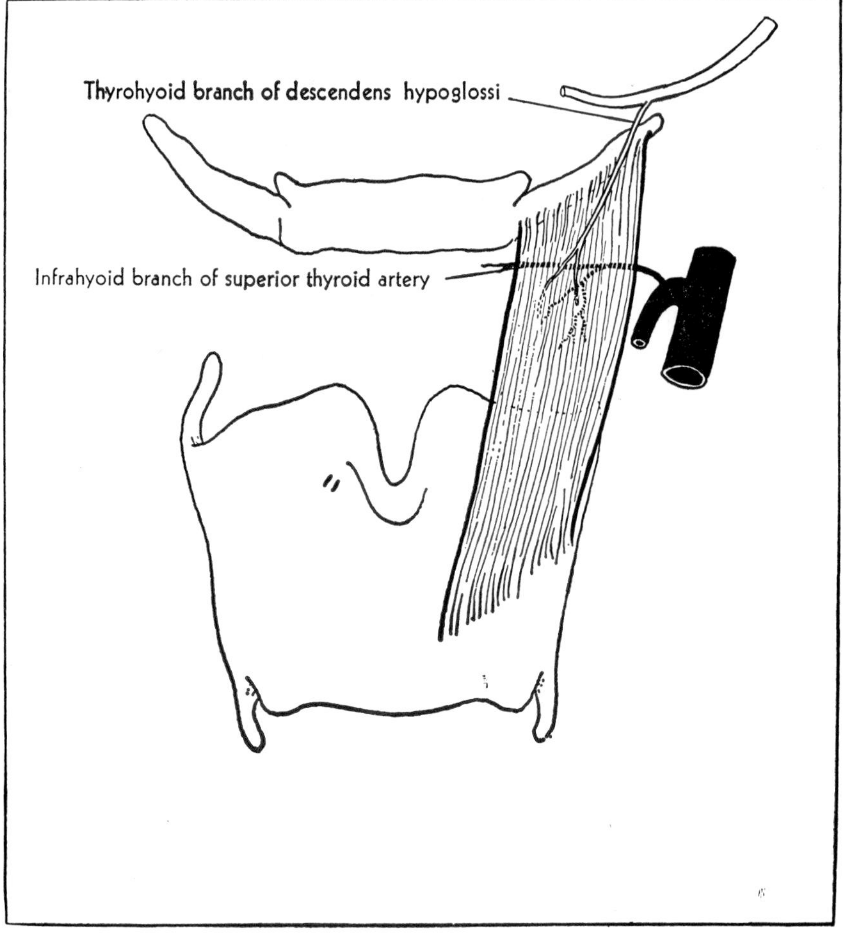

ORIGIN: Oblique line on lamina of thyroid cartilage
INSERTION: Lower border of body and greater horn of hyoid bone
FUNCTION: Depresses larynx and hyoid bone, elevates thyroid cartilage
NERVE: Thyrohyoid branch of CI through descendens hypoglossi
ARTERY: Infrahyoid branch of superior thyroid

REFERENCES:	GRAY	GRANT'S ATLAS
Muscle	399	529, 541
Nerve	399, 946, 955, 958	542, 546, 662
Artery	581	529, 542, 548

OMOHYOIDEUS

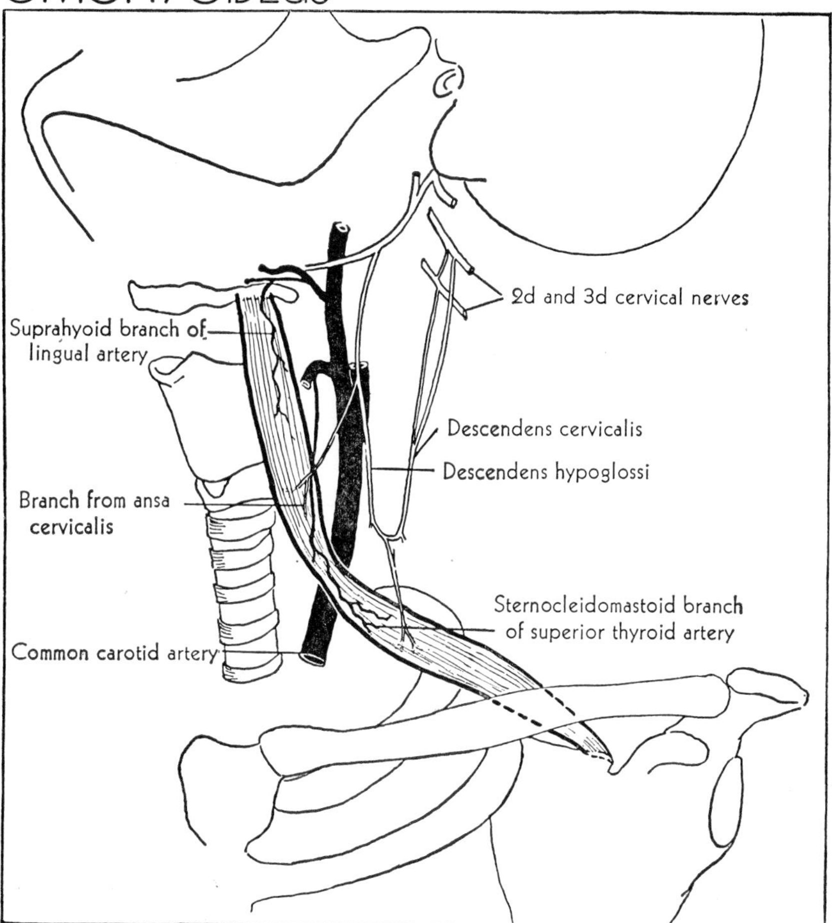

ORIGIN: Inferior belly from upper border of scapula and suprascapular ligament, ending in tendon under sternocleidomastoid muscle; superior belly extends upward from this tendon

INSERTION: Lower border of body of hyoid bone

FUNCTION: Steadies hyoid bone, depresses and retracts hyoid and larynx

NERVE: Ansa cervicalis formed by medial branches of 2d and 3d cervical and descending ramus of hypoglossal from CI

ARTERY: Suprahyoid branch of lingual; sternocleidomastoid branch of superior thyroid

REFERENCES:	GRAY	GRANT'S ATLAS
Muscle	399	25, 474, 528, 542
Nerve	399, 946, 955, 958	542, 662
Artery	581, 582	542

LONGUS COLLI (L. cervicis)

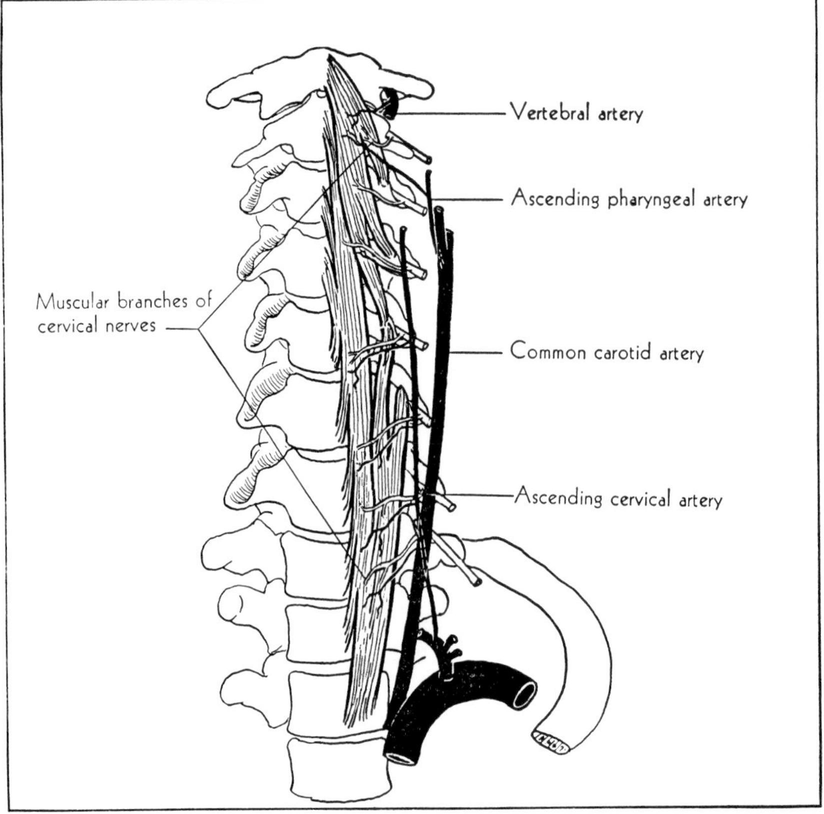

ORIGIN: a) Vertical portion from bodies of 1st 3 thoracic and last 3 cervical vertebrae; b) inferior oblique portion from bodies of 1st 3 thoracic vertebrae; c) superior oblique portion from anterior tubercles of transverse processes of 3d, 4th and 5th cervical vertebrae

INSERTION: a) Vertical portion into bodies of 2d, 3d and 4th cervical vertebrae; b) inferior oblique portion on anterior tubercles of transverse processes of 5th and 6th cervical vertebrae; c) superior oblique portion on anterior tubercle of atlas

FUNCTION: Flexes and assists in rotating cervical vertebrae and head; acting singly flexes column laterally

NERVE: Branches of anterior primary rami of 2d to 8th cervical

ARTERY: Prevertebral branches of ascending pharyngeal; muscular branches of ascending cervical and vertebral

REFERENCES:	GRAY	GRANT'S ATLAS
Muscle	399	561
Nerve	400, 955, 958, 963	Not shown
Artery	581, 603, 605	536

LONGUS CAPITIS

ORIGIN: Anterior tubercles of transverse processes of 3d, 4th, 5th, 6th cervical vertebrae

INSERTION: Inferior surface of basilar part of occipital bone

FUNCTION: Flexes and assists in rotating cervical vertebrae and head

NERVE: Muscular branches of 1st, 2d, 3d, 4th cervical

ARTERY: Ascending cervical of inferior thyroid; prevertebral of ascending pharyngeal; muscular of vertebral

REFERENCES:	GRAY	GRANT'S ATLAS
Muscle	400	561
Nerve	400, 955, 958	Not shown
Artery	581, 603, 605	536

RECTUS CAPITIS ANTERIOR

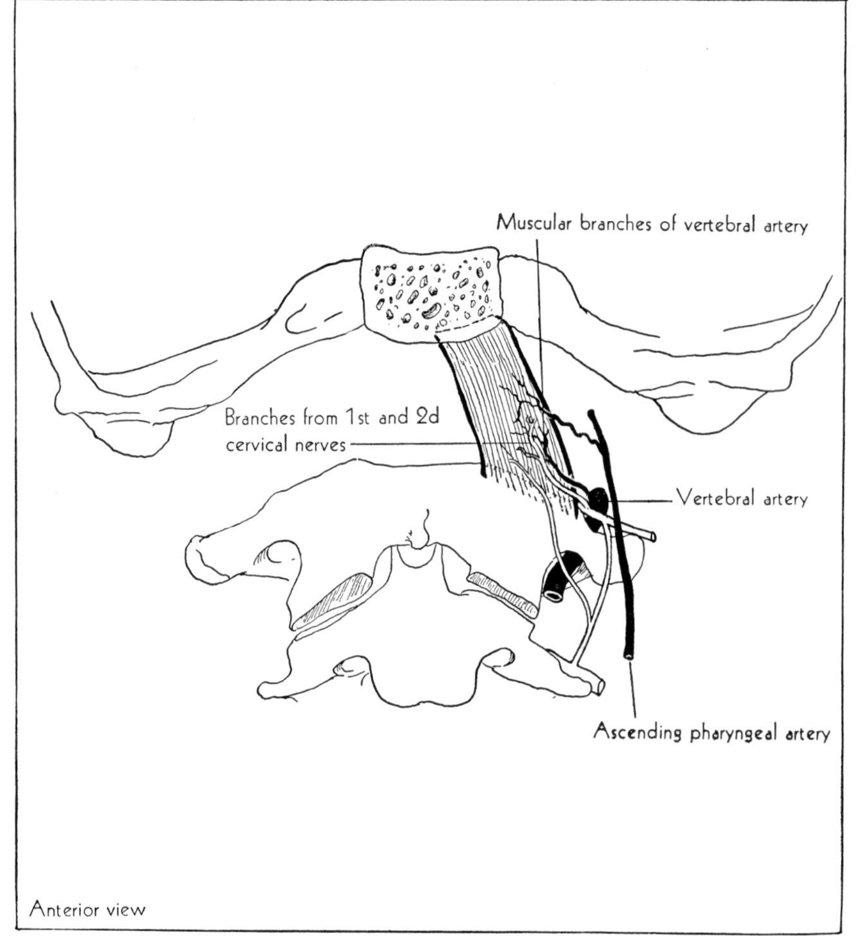

ORIGIN: Lateral mass of atlas
INSERTION: Base of occipital bone in front of foramen magnum
FUNCTION: Flexes and rotates head
NERVE: Muscular branches of 1st and 2d cervical
ARTERY: Muscular branches of vertebral; ascending pharyngeal

REFERENCES:	GRAY	GRANT'S ATLAS
Muscle	400	561
Nerve	400, 955, 958	561
Artery	581, 603	Not shown

RECTUS CAPITIS LATERALIS

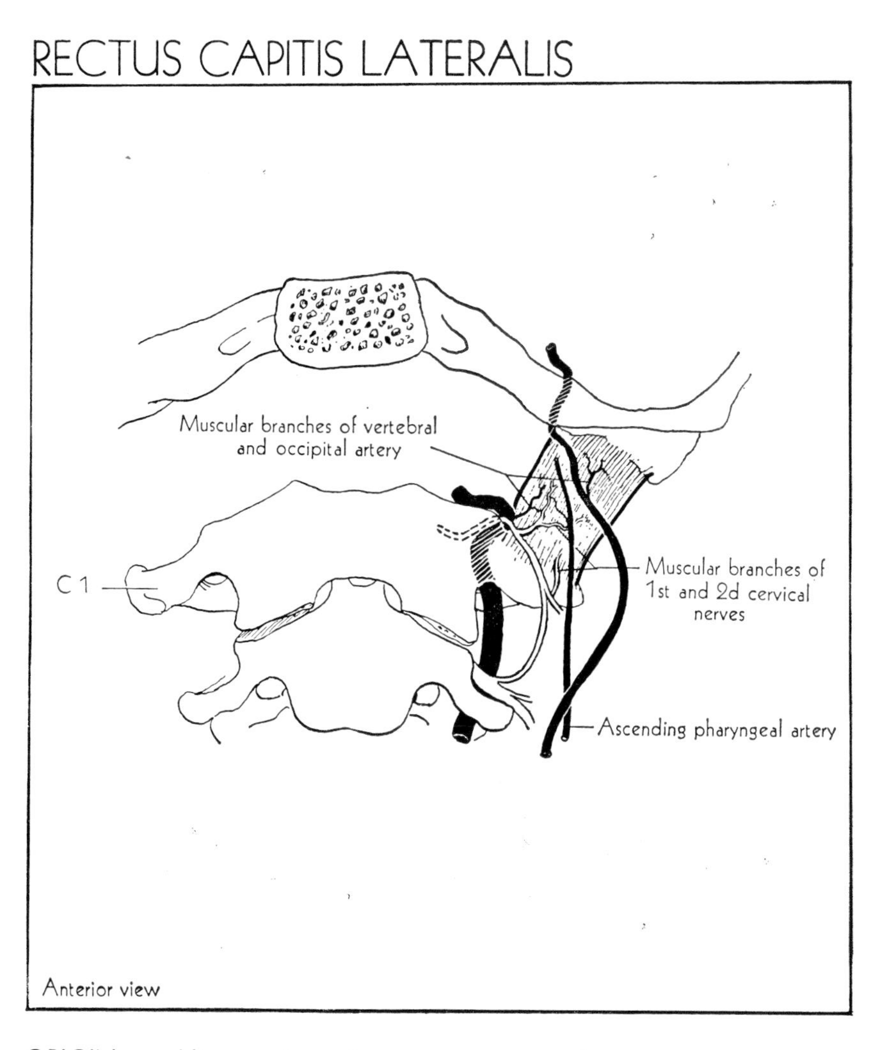

ORIGIN: Upper surface of transverse process of atlas
INSERTION: Inferior surface of jugular process of occipital bone
FUNCTION: Flexes head laterally
NERVE: Branches of anterior rami of 1st and 2d cervical
ARTERY: Muscular branches of vertebral; occipital; ascending pharyngeal

REFERENCES:	GRAY	GRANT'S ATLAS
Muscle	400	495, 561
Nerve	400, 955, 958	561
Artery	581, 584, 603	561

SCALENUS ANTERIOR

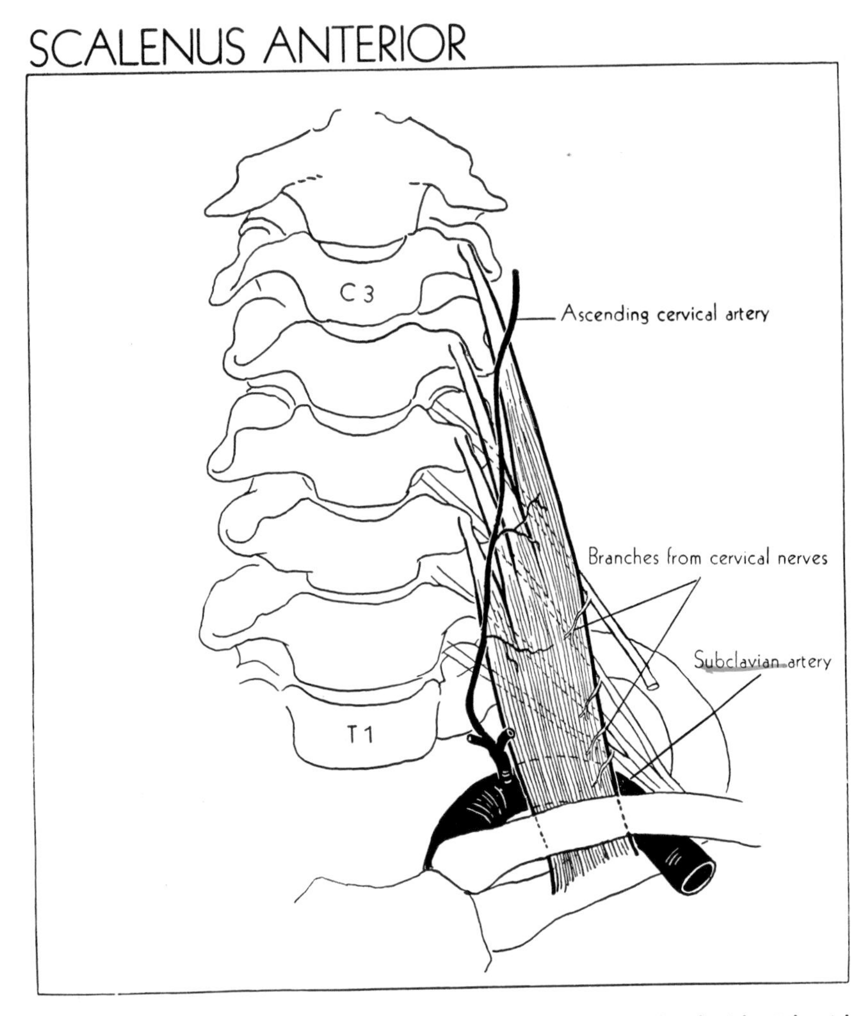

ORIGIN: Anterior tubercles of transverse processes of 3d, 4th, 5th, 6th cervical vertebrae

INSERTION: Scalene tubercle and ridge on upper surface of 1st rib

FUNCTION: Acting from above elevates 1st rib; from below, flexes and rotates cervical column

NERVE: Anterior branches of 5th to 8th cervical

ARTERY: Ascending cervical branch of inferior thyroid

REFERENCES: GRAY		GRANT'S ATLAS
Muscle	400	474, 531, 536, 561
Nerve	400, 961, 963	Not shown
Artery	605	531, 536, 561

SCALENUS MEDIUS

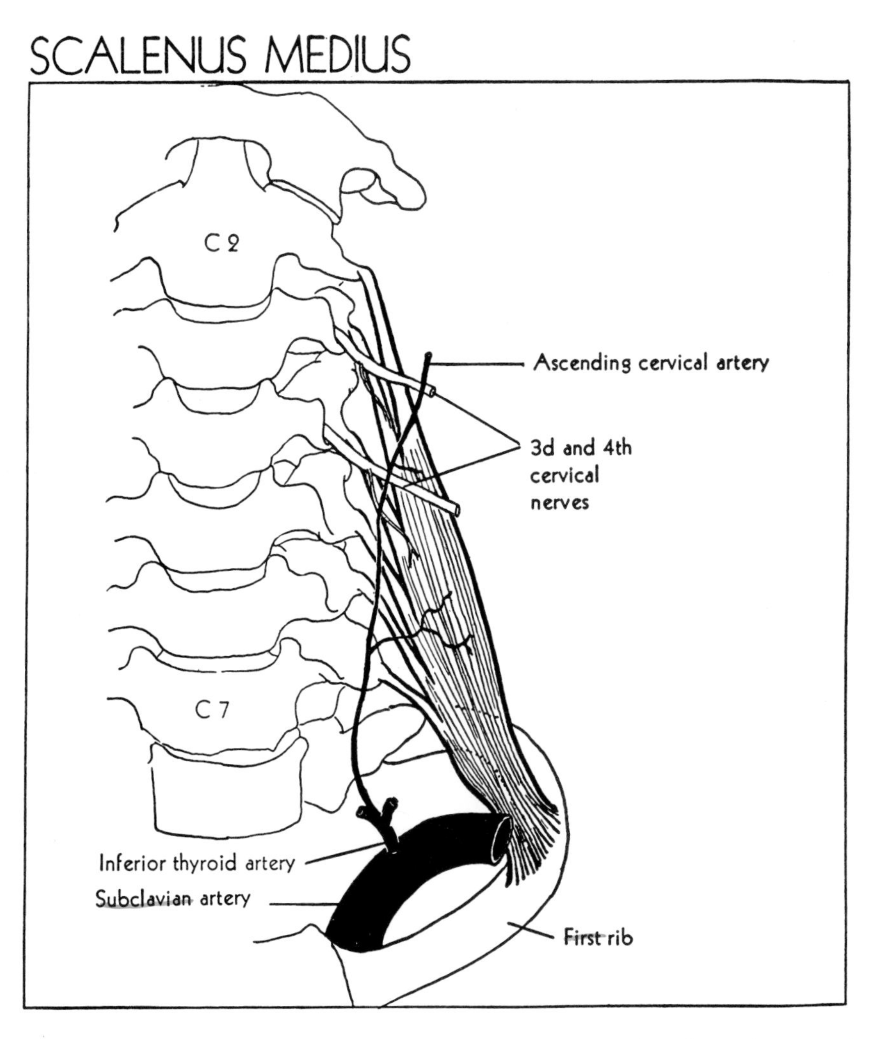

ORIGIN: Posterior tubercles of transverse processes of 2nd to 7th cervical vertebrae

INSERTION: Upper surface of 1st rib behind subclavian groove

FUNCTION: Flexes and assists in rotating cervical vertebrae and head; flexes cervical column laterally, acting from above elevates 1st rib; inspiratory

NERVE: Posterior branches of anterior primary rami of 3d and 4th cervical, lateral muscular branches of 3d and 4th cervical

ARTERY: Muscular branches of ascending cervical

REFERENCES:

	GRAY	GRANT'S ATLAS
Muscle	400	473-475, 561
Nerve	400, 955, 961, 963	Not shown
Artery	605	Not shown

SCALENUS POSTERIOR

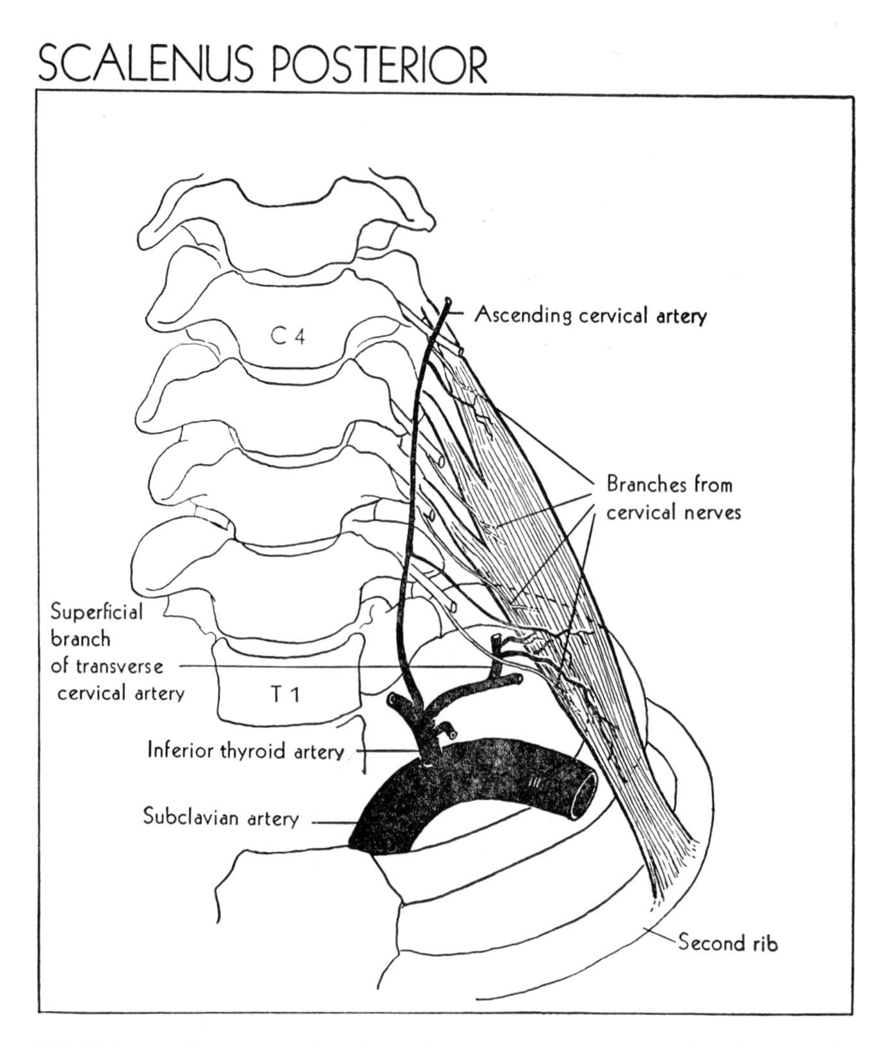

ORIGIN: Posterior tubercles of transverse processes of 4th, 5th, 6th cervical vertebrae

INSERTION: Outer surface of 2d rib behind attachment of serratus anterior

FUNCTION: Flexes and assists in rotating cervical vertebrae and head; acting singly, flexes cervical column laterally; acting from above, elevates 2d rib; inspiratory

NERVE: Posterior branches of anterior primary rami of lower 4 cervical, lateral muscular branches of 3d and 4th cervical

ARTERY: Muscular branches of ascending cervical division of inferior thyroid, superficial branch of transverse cervical

REFERENCES:

	GRAY	GRANT'S ATLAS
Muscle	400	474, 475
Nerve	400, 961, 963	Not shown
Artery	605, 606	474, 475

SPLENIUS CAPITIS AND CERVICIS

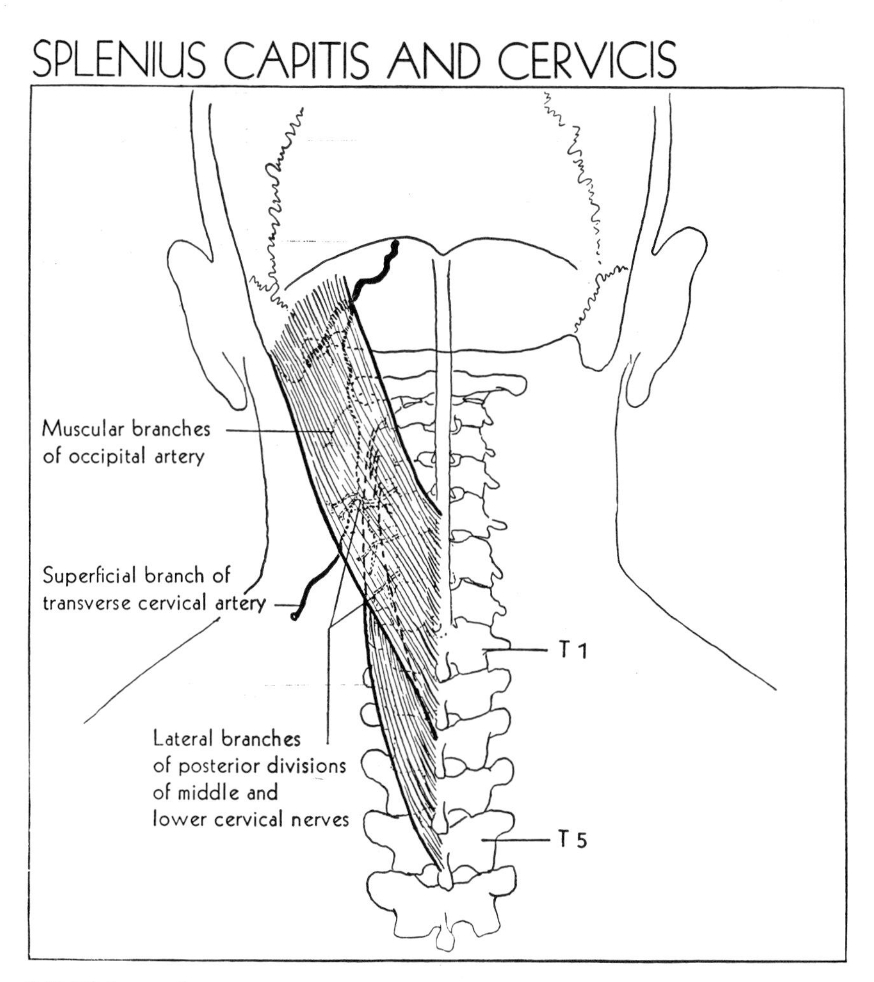

ORIGIN: Sp. capitis from lower half of ligamentum nuchae and spine of 7th cervical and upper 3 or 4 thoracic vertebrae; Sp. cervicis from spines of 3rd to 6th thoracic vertebrae

INSERTION: Sp. capitis into mastoid process of temporal bone and lateral part of superior nuchal line; Sp. cervicis into posterior tubercles of transverse processes of upper 3 or 4 cervical vertebrae

FUNCTION: Together they extend, laterally flex head and neck and rotate head slightly

NERVE: Lateral branches of posterior primary rami of middle and lower cervical

ARTERY: Muscular and descending branches of occipital, superficial branch of transverse cervical

REFERENCES:	GRAY	GRANT'S ATLAS
Muscle	403	473, 479, 494
Nerve	403, 949	Not shown
Artery	585, 606	491

ERECTOR SPINAE (Sacrospinalis)

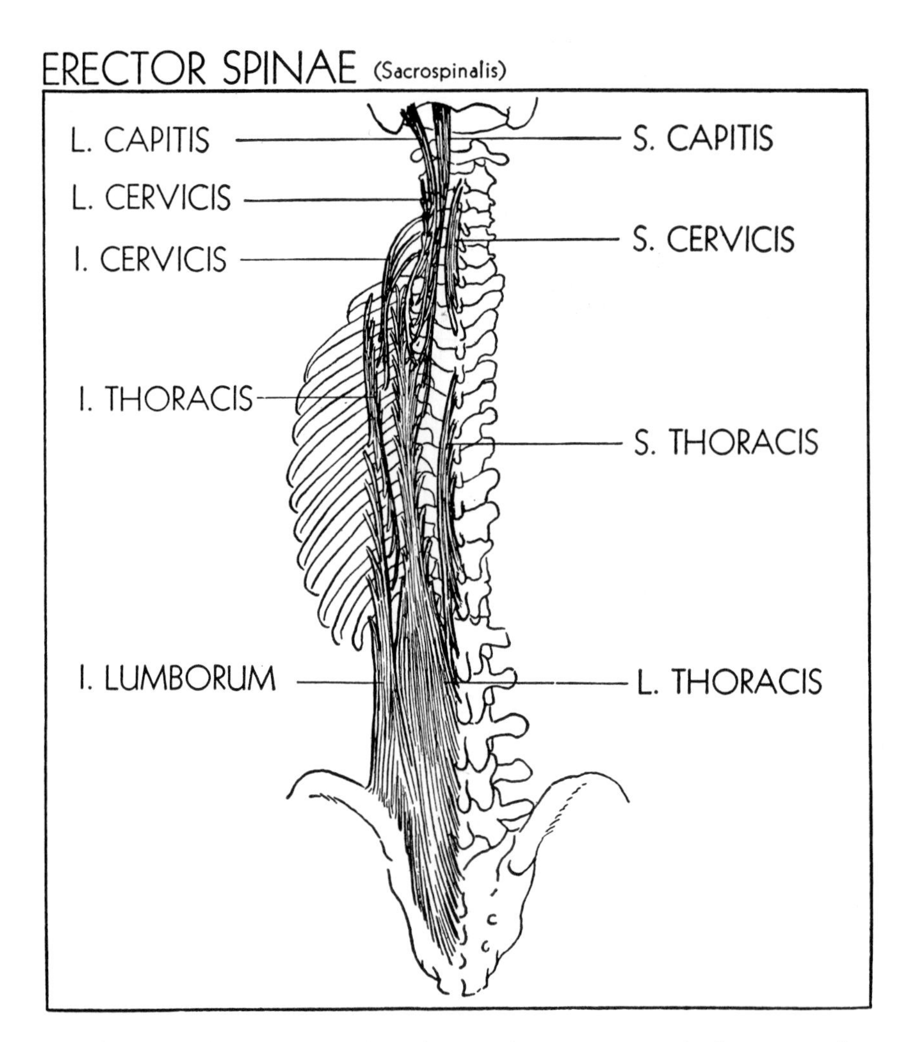

ORIGIN and INSERTION: This muscle is "an elongated mass composed of separate slips extending from sacrum to skull" (Cunningham). It arises from the sacrum, crest of ilium, spines of 11th, 12th thoracic and lumbar vertebrae. It separates into a lateral iliocostalis, an intermediate longissimus, and a medial spinalis column described on pages 53, 54, 55.

FUNCTION: Extends, laterally flexes and rotates vertebral column

NERVE: Posterior primary rami of spinal nerves according to their situation

ARTERY: Muscular branches of vertebral; Deep cervical; Posterior rami of intercostals and lumbars

REFERENCES:	GRAY	GRANT'S ATLAS
Muscle	405	178, 479
Nerve	405, 949-953	Not shown
Artery	603, 610, 627, 641	Not shown

ILIOCOSTALIS LUMBORUM, I. THORACIS, I. CERVICIS

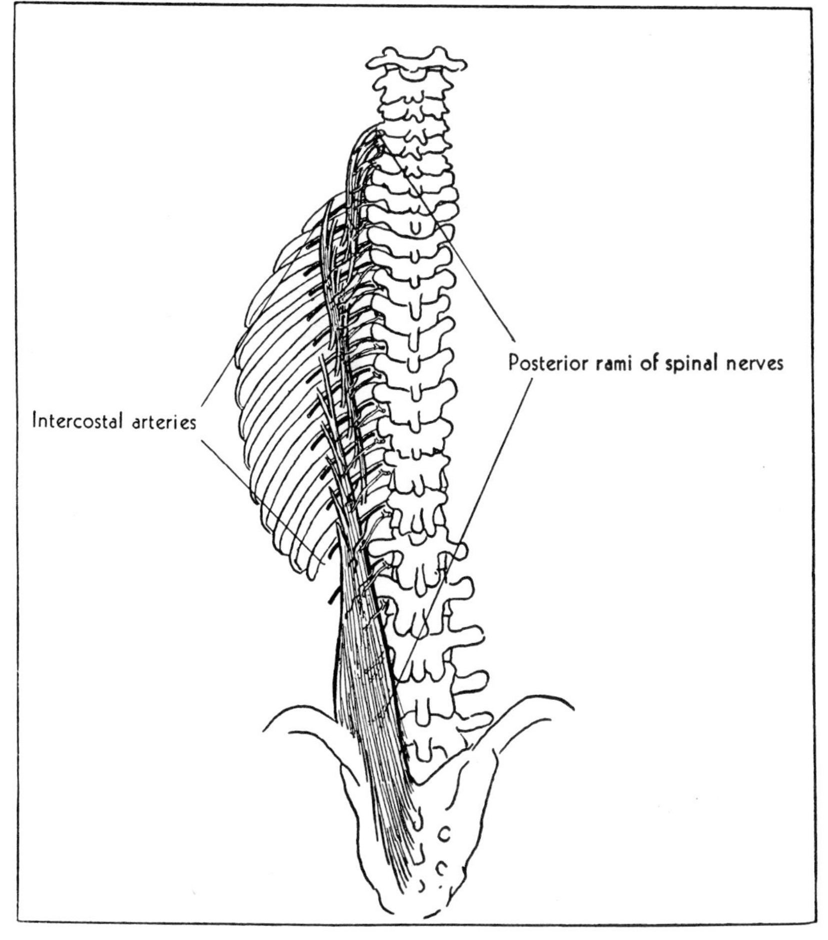

ORIGIN: a) I. lumborum, body of sacrospinalis in lumbar region; b) I thoracis, angles of lower 6 ribs medial to insertion of I. lumborum; c) I. cervicis, angles of 3d to 6th ribs

INSERTION: a) I. lumborum, lower borders of angles of lower 6 or 7 ribs, b) I. thoracis, upper borders of angles of upper 6 ribs; c) I. cervicis, transverse processes of 4th to 6th cervical vertebrae

FUNCTION: Extension, lateral flexion and rotation of column; lateral movement of pelvis

NERVE: Posterior primary rami of spinal

ARTERY: Posterior rami of intercostals and lumbars

REFERENCES: GRAY		GRANT'S ATLAS
Muscle	405	479, 481
Nerve	405, 949-953	Not shown
Artery	627, 641	Not shown

LONGISSIMUS THORACIS, L. CERVICIS, L. CAPITIS

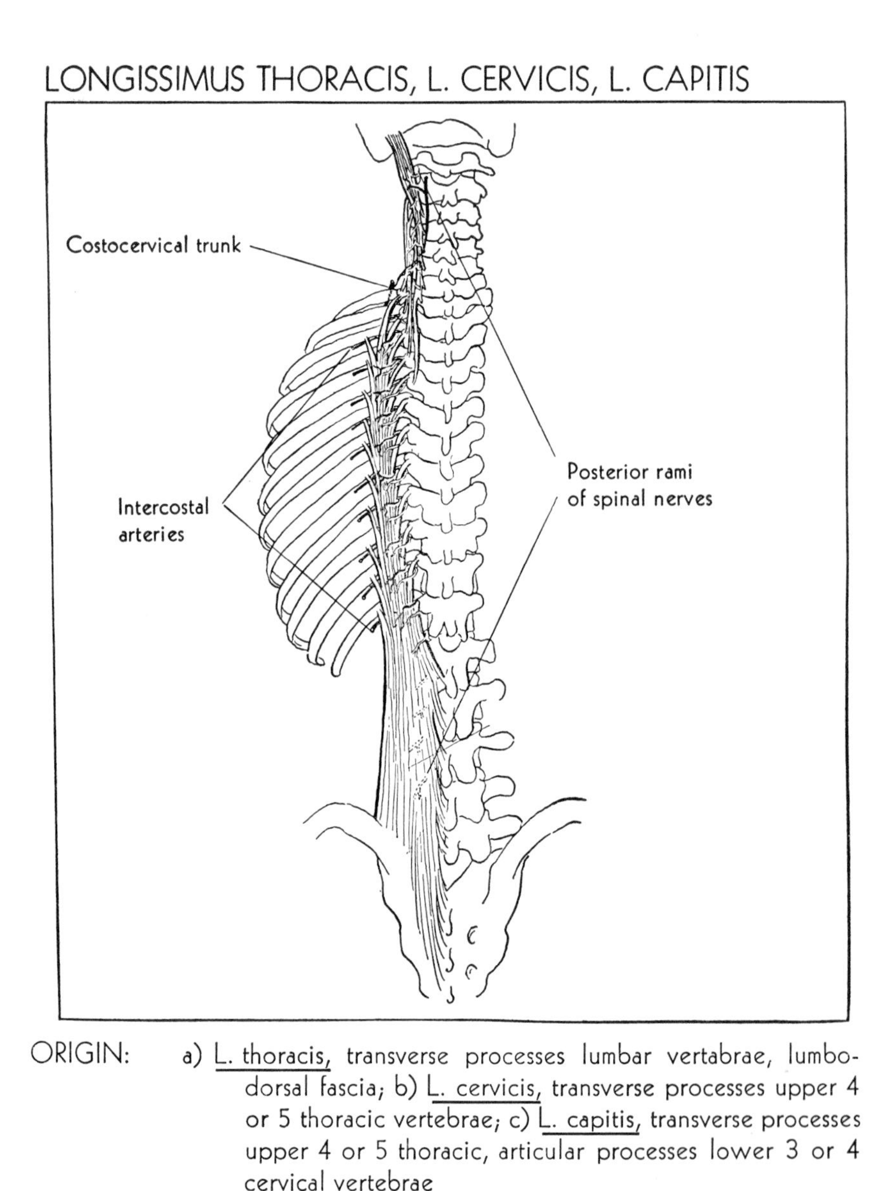

ORIGIN: a) L. thoracis, transverse processes lumbar vertabrae, lumbodorsal fascia; b) L. cervicis, transverse processes upper 4 or 5 thoracic vertebrae; c) L. capitis, transverse processes upper 4 or 5 thoracic, articular processes lower 3 or 4 cervical vertebrae

INSERTION: a) L. thoracis, transverse processes thoracic vertebrae, lower 9 or 10 ribs proximal to angles; b) L. cervicis, transverse processes 2d to 6th cervical vertebrae; c) L. capitis, posterior margin mastoid process

FUNCTION: Extension, lateral flexion and rotation of column; lateral movement of pelvis

NERVE: Posterior primary rami of spinal

ARTERY: Posterior rami of intercostals and lumbars; muscular branches of occipital: deep cervical branch of costocervical trunk

REFERENCES:

	GRAY	GRANT'S ATLAS
Muscle	405	479, 481, 491
Nerve	405, 406, 949-953	Not shown
Artery	585, 610, 627, 641	491

SPINALIS THORACIS, S. CERVICIS, S. CAPITIS

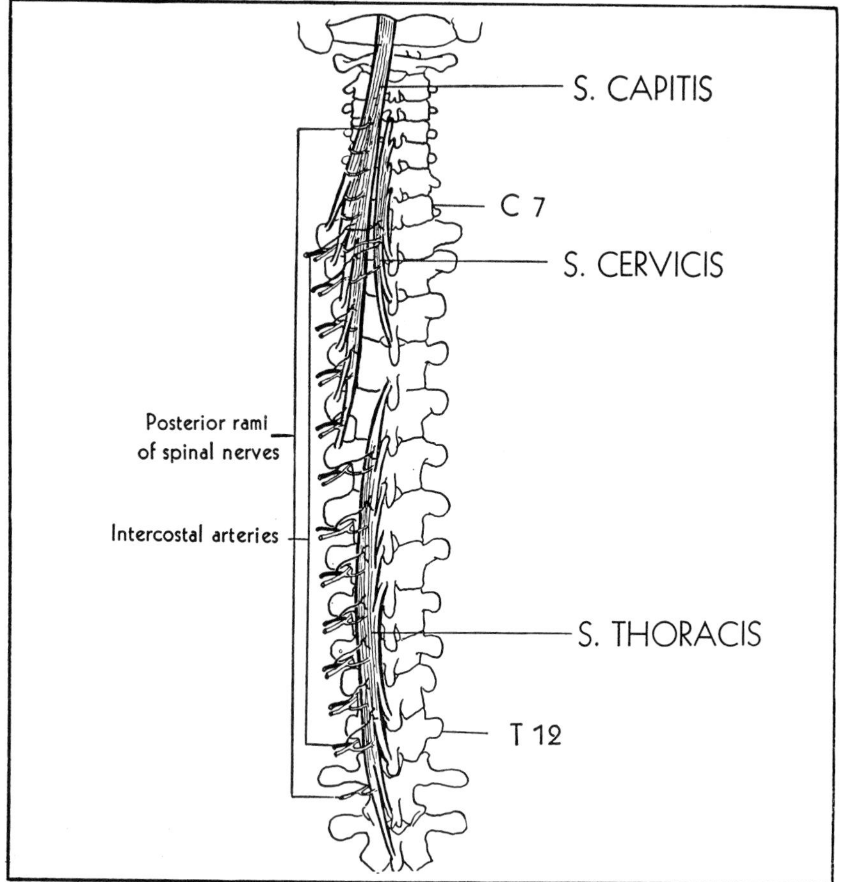

ORIGIN: a) S. thoracis, spines 1st, 2d lumbar, 11th, 12th thoracic vertebrae; b) S. cervicis, spines 1st, 2d thoracic, 7th cervical vertebrae, lower part of ligamentum nuchae; c) S. capitis, transverse processes, upper 6 or 7 thoracic, 7th cervical, articular processes 4th to 6th cervical vertebrae

INSERTION: a) S. thoracis, spines upper 4 to 8 thoracic vertebrae; b) S. cervicis, spines of 2d and sometimes of 3d, 4th, cervical vertebrae; c) S. capitis, occipital bone between superior and inferior nuchal lines

FUNCTION: Extension, lateral flexion and rotation of column; lateral movement of pelvis

NERVE: Posterior primary rami of spinal

ARTERY: Posterior rami of intercostals; deep cervical branch of costocervical trunk

REFERENCES:	GRAY	GRANT'S ATLAS
Muscle	406	479
Nerve	406, 949-953	491
Artery	610, 627	Not shown

SEMISPINALIS THORACIS, S. s. CERVICIS, S. s. CAPITIS

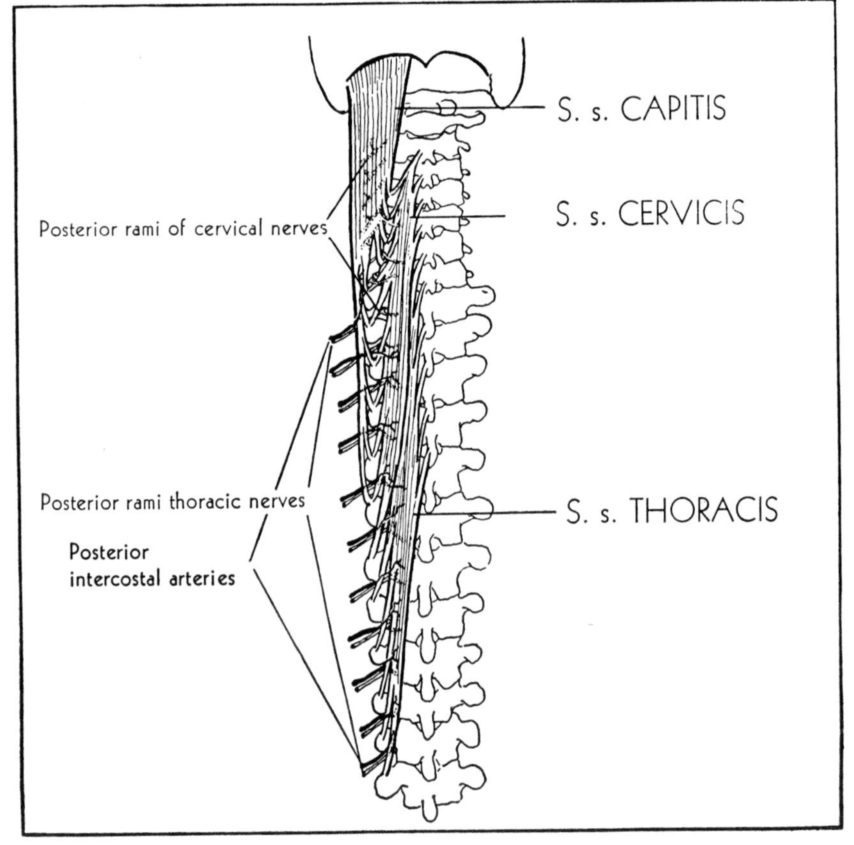

ORIGIN: a) Ss. thoracis, transverse processes lower 6 thoracic vertebrae; b) Ss. cervicis, transverse processes upper 6 thoracic, articular processes lower 4 cervical vertebrae; c) Ss. capitis, transverse processes upper 6 thoracic, 7th cervical, articular processes of 4th to 6th cervical

INSERTION: a) Ss. thoracis, spines of first 4 thoracic, last 2 cervical; b) Ss. cervicis, spines of 2d to 5th cervical vertebrae; c) Ss. capitis, occipital bone, medial impression between superior and inferior nuchal lines

FUNCTION: Extension and lateral flexion of column; extension of head, ribs and pelvis

NERVE: Ss. thoracis, medial branches posterior primary rami of upper 6 thoracic; Ss. cervicis, posterior rami of lower 3 cervical; Ss. capitis, 1st to 6th cervical

ARTERY: Muscular branches of posterior intercostals, descending branch of occipital; deep cervical branch of costocervical trunk

REFERENCES:	GRAY	GRANT'S ATLAS
Muscle	406	479, 490, 494
Nerve	406, 949-953	490, 491
Artery	585, 610, 627	491, 492

MULTIFIDUS

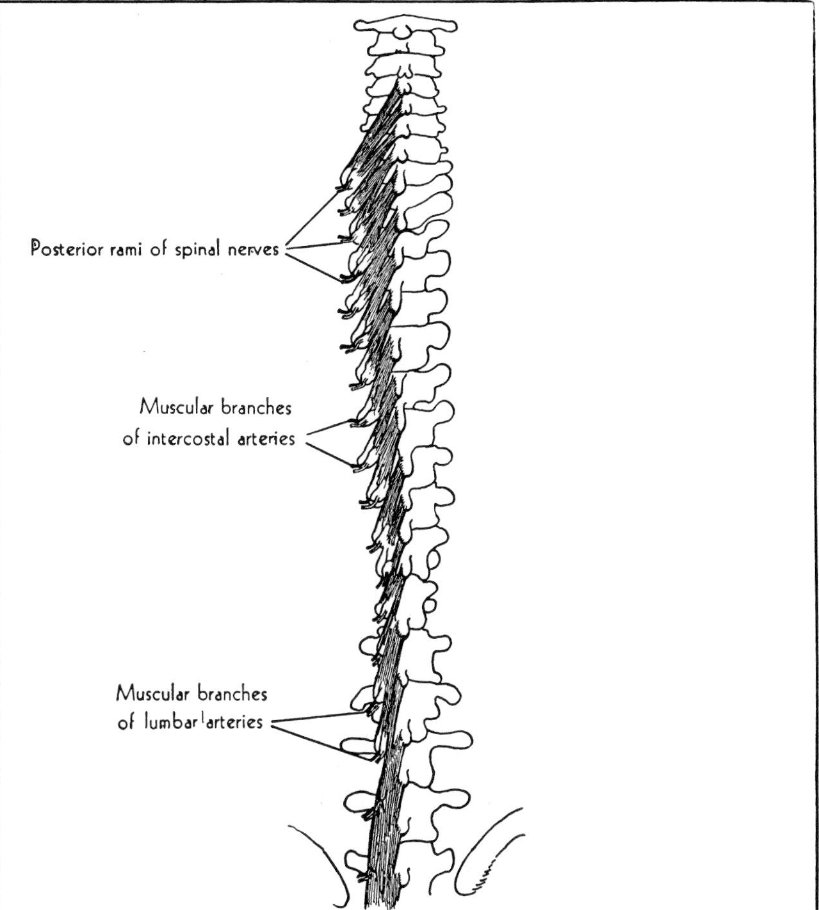

ORIGIN: Back of sacrum, posterior sacro-iliac ligament, mammillary processes of lumbar, transverse processes of thoracic, articular processes of lower 4 cervical vertebrae

INSERTION: Spine of vertebra above vertebra of origin

FUNCTION: Aid in extension, lateral flexion, and rotation of column; extension and lateral movement of pelvis

NERVE: Posterior primary rami of all spinal nerves

ARTERY: Medial muscular branches of posterior intercostals and lumbars; deep cervical branch of costocervical trunk

REFERENCES:	GRAY	GRANT'S ATLAS
Muscle	406	480, 481, 493
Nerve	407, 949-953	493
Artery	610, 627, 641	Not shown

ROTATORES

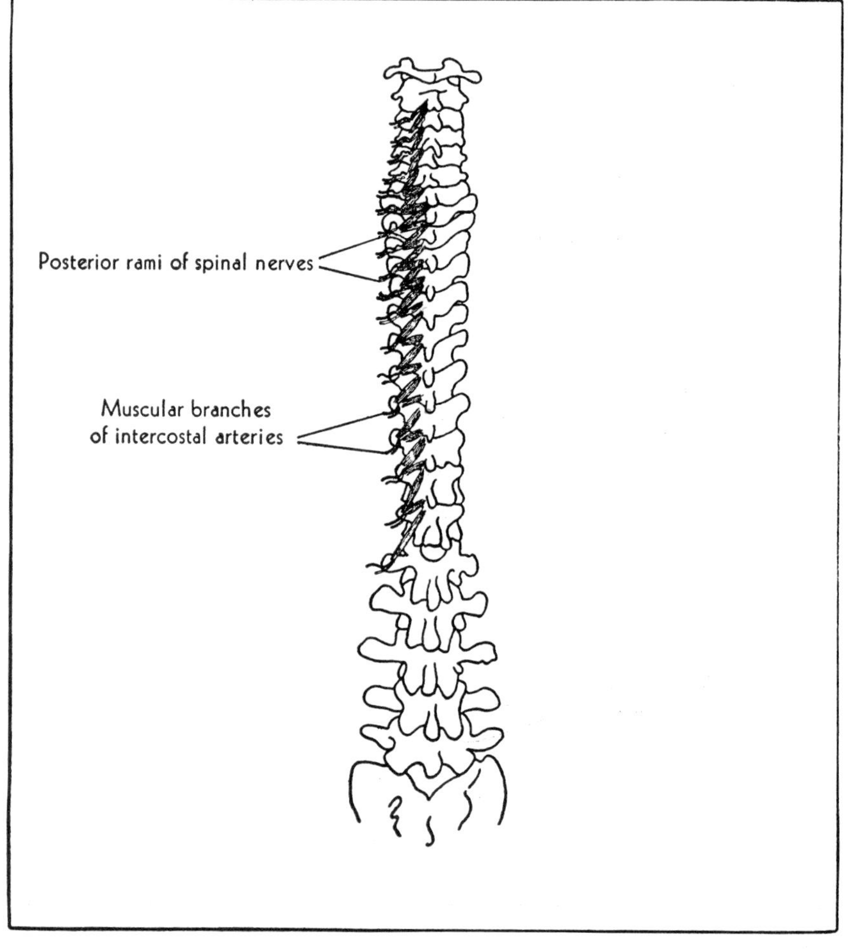

ORIGIN: Lying deep to the multifidus they form 11 pairs of small muscles, each arises from the transverse process of one thoracic vertebra

INSERTION: Into lamina of vertebra directly above vertebra of origin

FUNCTION: Assist in rotating vertebral column

NERVE: Posterior primary rami of thoracic nerves

ARTERY: Muscular branches of posterior intercostals

REFERENCES:

	GRAY	GRANT'S ATLAS
Muscle	407	411, 493
Nerve	407, 949-952	411
Artery	627	411

INTERSPINALES

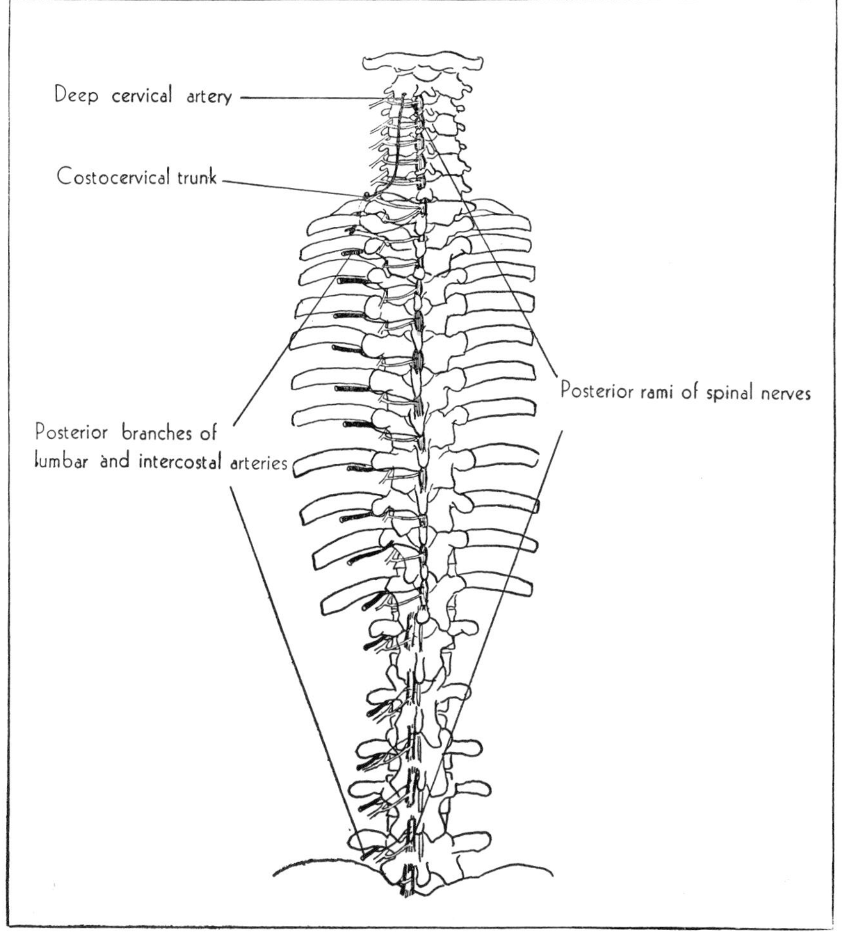

ORIGIN and INSERTION: Short muscular fasciculi connecting spines of vertebrae from lumbar region to 2d cervical vertebra; not always present in thoracic region; frequently double in cervical region

FUNCTION: Aid in extension of column

NERVE: Posterior primary rami of spinal nerves

ARTERY: Muscular branches of posterior intercostals, lumbar, deep cervical branch of costocervical trunk

REFERENCES:	GRAY	GRANT'S ATLAS
Muscle	407	481, 493, 538
Nerve	407, 949-953	538
Artery	610, 627, 641	Not shown

INTERTRANSVERSARII

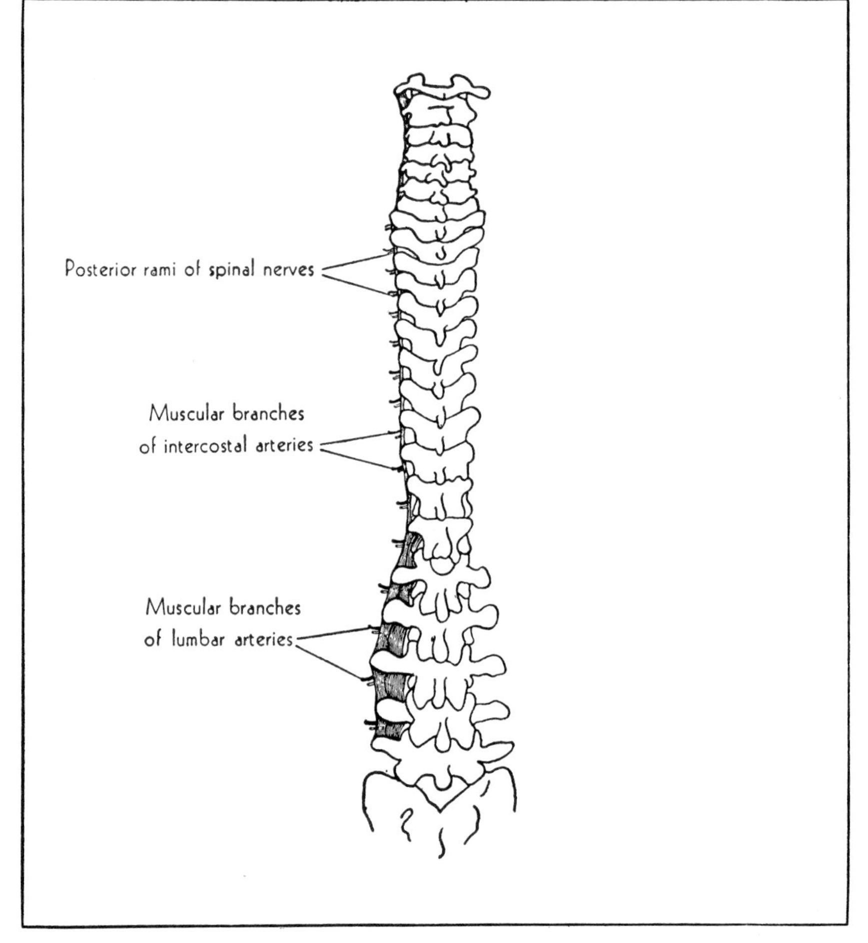

ORIGIN and INSERTION: Slender muscular slips extending between adjacent transverse processes of vertebrae, best represented in cervical and lumbar region

FUNCTION: Probably assist in lateral flexion of the vertebral column

NERVE: Posterior primary rami of spinal nerves, with exception of lateral intertransversarii in lumbar and lower thoracic region and anterior and posterior intertransversarii in cervical region; these are supplied by anterior primary rami of spinal nerves

ARTERY: Deep cervical branch of costocervical trunk. Muscular branches of posterior intercostals and lumbars

REFERENCES:	GRAY	GRANT'S ATLAS
Muscle	407	480, 481, 538, 561
Nerve	408, 949-953, 955, 982-983	538
Artery	**610, 627, 641**	**Not shown**

RECTUS CAPITIS POSTERIOR MAJOR

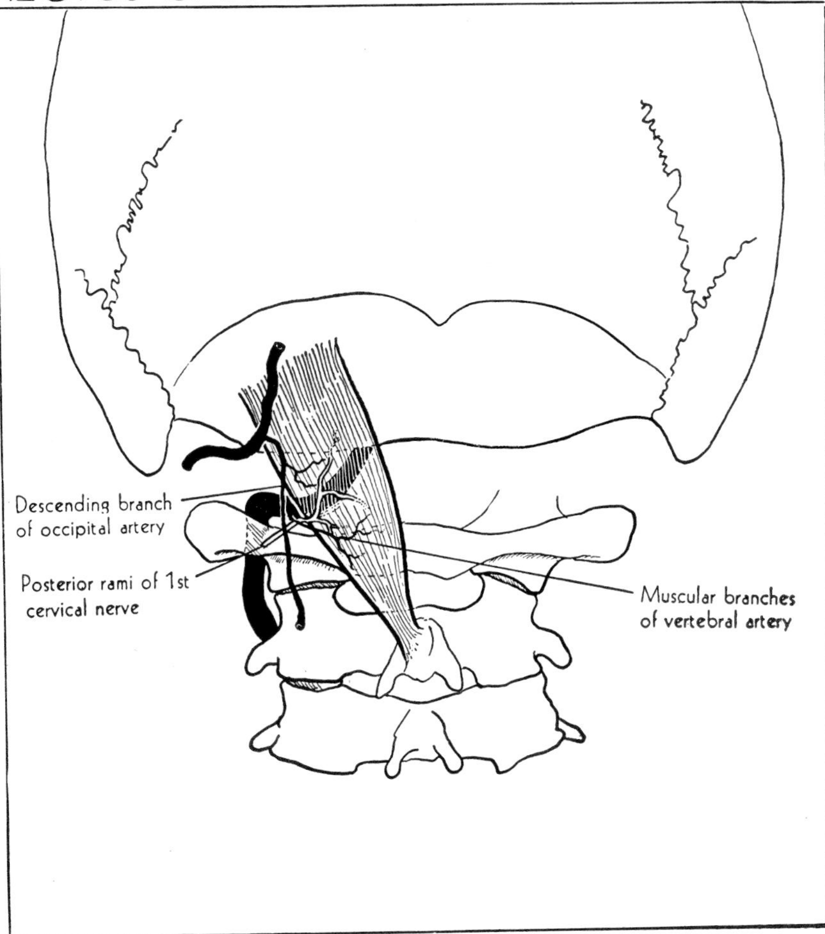

ORIGIN: Spine of axis
INSERTION: Lateral part of inferior nuchal line of occipital bone
FUNCTION: Extension, lateral flexion and rotation of head
NERVE: Muscular branches of posterior primary ramus of 1st cervical (Suboccipital)
ARTERY: Muscular branches of vertebral, descending branch of occipital

REFERENCES:	GRAY	GRANT'S ATLAS
Muscle	408	490, 492, 494
Nerve	409, 949	491, 492
Artery	585, 603	491

RECTUS CAPITIS POSTERIOR MINOR

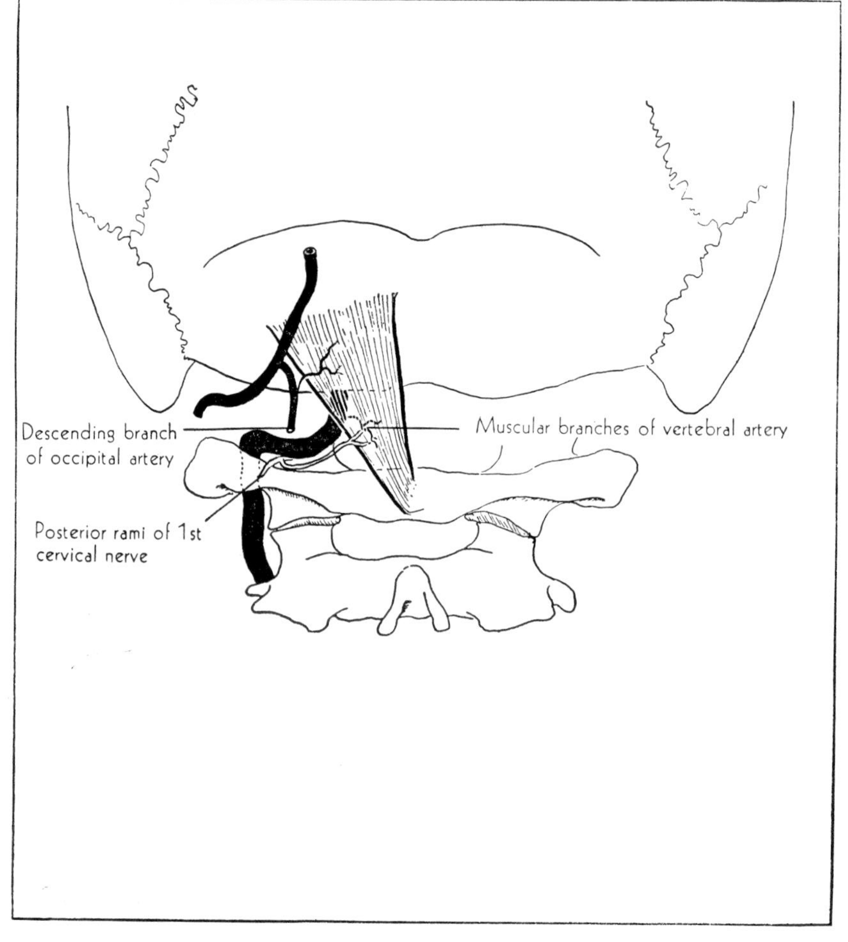

ORIGIN: Posterior tubercle of the atlas
INSERTION: Occipital bone below inferior nuchal line
FUNCTION: Extension and lateral flexion of head
NERVE: Muscular branches of posterior primary ramus of 1st cervical (Suboccipital)
ARTERY: Muscular branches of vertebral, descending branch of occipital

REFERENCES: GRAY		GRANT'S ATLAS
Muscle	409	490, 492
Nerve	409, 949	491, 492
Artery	585, 603	491

OBLIQUUS CAPITIS INFERIOR

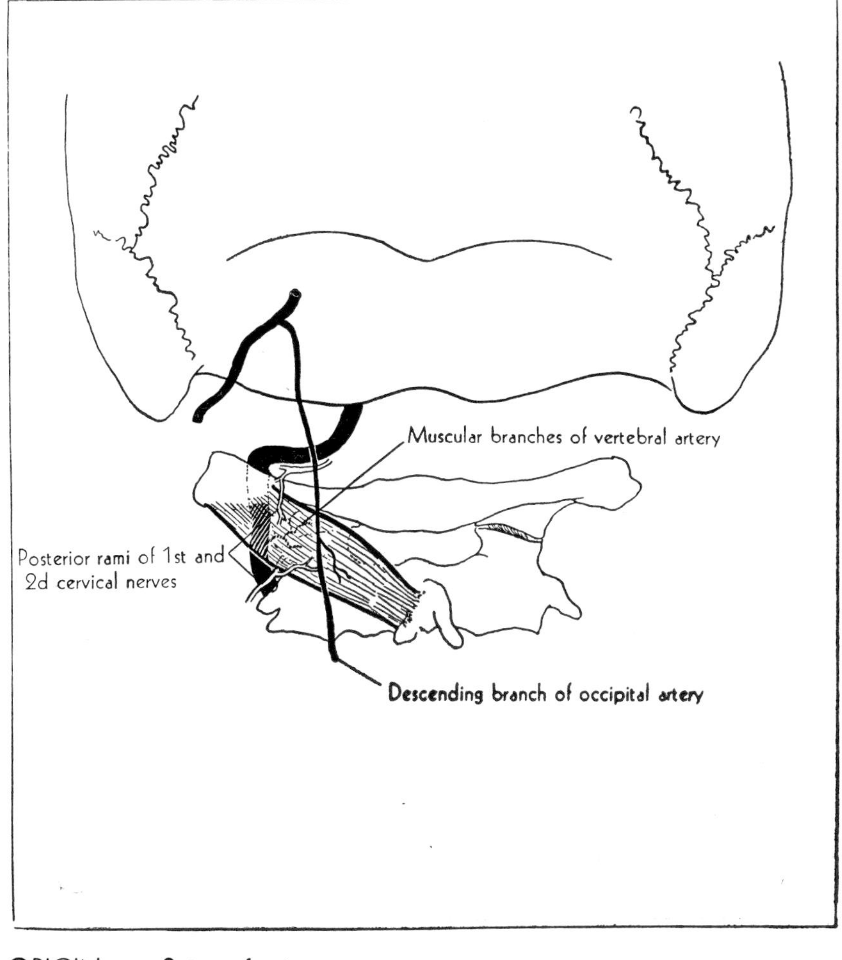

ORIGIN: Spine of axis
INSERTION: Transverse process of atlas
FUNCTION: Rotates atlas and skull around odontoid process of axis
NERVE: Muscular branches of posterior primary rami of 1st and 2d cervical
ARTERY: Muscular branches of vertebral, descending branch of occipital

REFERENCES:

	GRAY	GRANT'S ATLAS
Muscle	409	490, 492, 494
Nerve	409, 949	490-492
Artery	585, 603	Not shown

OBLIQUUS CAPITIS SUPERIOR

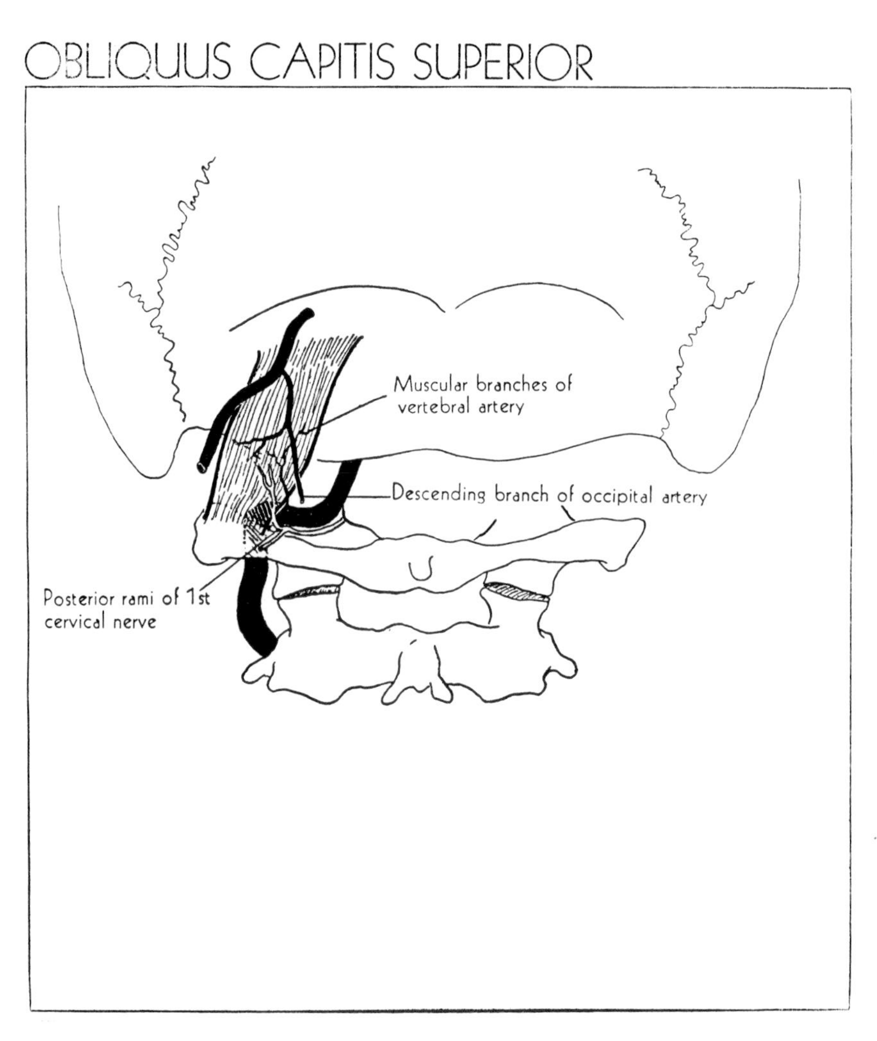

ORIGIN: Transverse process of atlas
INSERTION: Occipital bone above inferior nuchal line
FUNCTION: Extension and lateral rotation of head
NERVE: Muscular branches of posterior primary ramus of 1st cervical
ARTERY: Muscular branches of vertebral, descending branch of occipital

REFERENCES:	GRAY	GRANT'S ATLAS
Muscle	409	490, 492
Nerve	409, 949	490-492
Artery	585, 603	490, 491

INTERCOSTALES EXTERNI

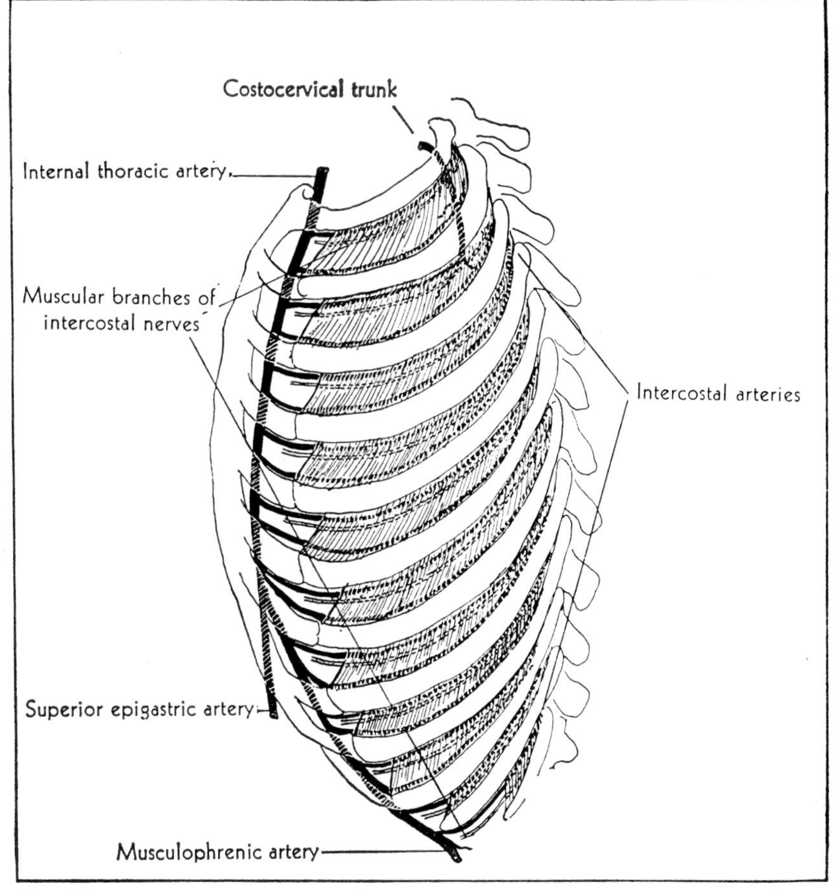

ORIGIN: 11 on either side, each arises from lower border of a rib, beginning at the tubercle behind to rib cartilages in front; anteriorly they terminate in intercostal membranes which continue to the sternum

INSERTION: Upper border of rib below the rib of origin

FUNCTION: Support intercostal spaces in inspiration and expiration; elevate ribs in inspiration

NERVE: Muscular branches of intercostals

ARTERY: Posterior intercostals, collateral branches of posterior intercostals, costocervical trunk, anterior intercostal branches of internal thoracic; musculophrenic

REFERENCES:	GRAY	GRANT'S ATLAS
Muscle	410	406, 408
Nerve	410, 979-982	406-409
Artery	609, 610, 627	406

INTERCOSTALES INTERNI

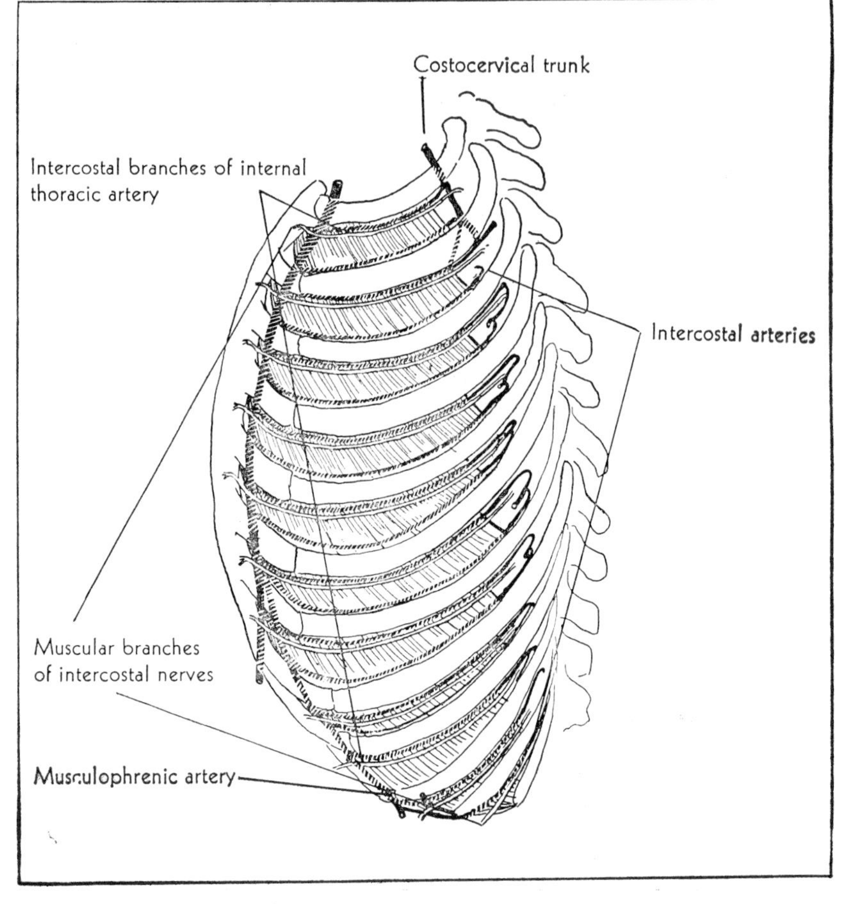

ORIGIN: 11 on either side, from upper surface of costal cartilages and ribs, they extend from sternum to angles of ribs

INSERTION: Costal cartilage and edge of costal groove of rib above rib of origin

FUNCTION: Prevent pushing out or drawing in of intercostal spaces in inspiration and expiration; lower the ribs in forced expiration

NERVE: Muscular branches of intercostals

ARTERY: Muscular branches of anterior intercostal, muscular branches of posterior intercostal, intercostal branches of internal thoracic and musculo-phrenic, costocervical trunk

REFERENCES:	GRAY	GRANT'S ATLAS
Muscle	410	405, 408
Nerve	411, 979-982	406-409
Artery	609, 610, 627	407

SUBCOSTALES

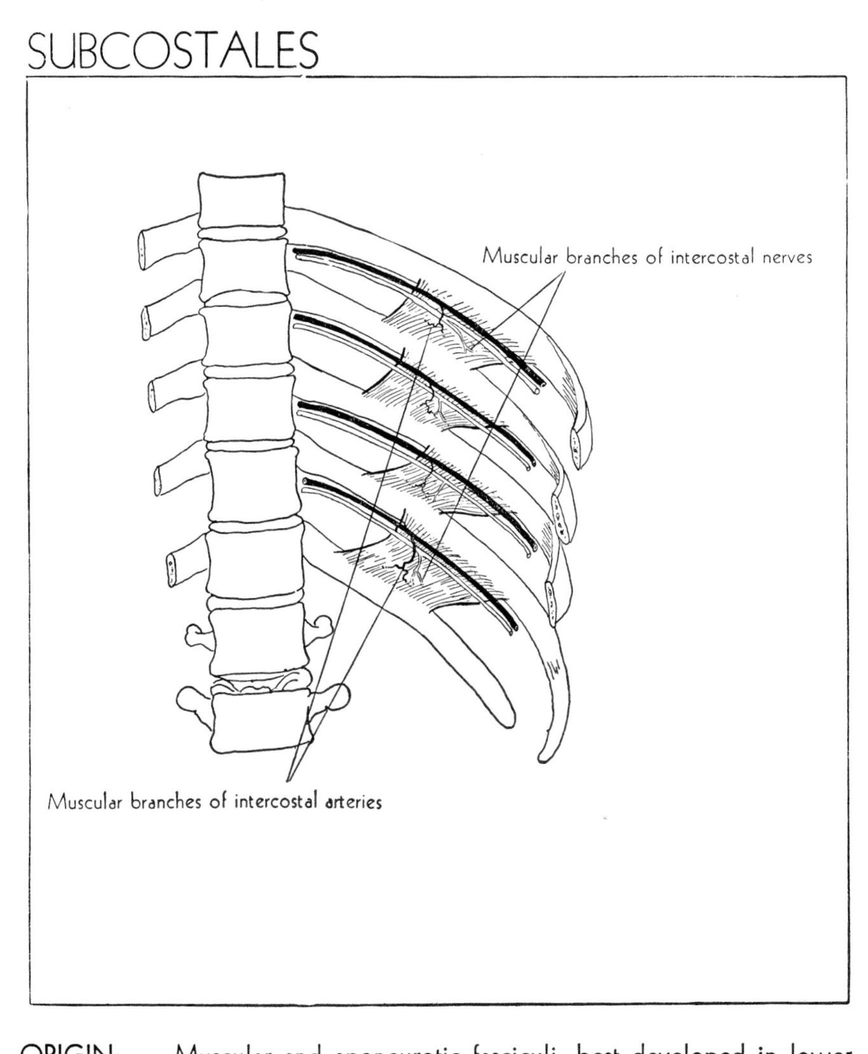

ORIGIN: Muscular and aponeurotic fasciculi, best developed in lower part of thorax; they arise from inner surface of ribs near their angles

INSERTION: Upper border of rib below rib of origin

FUNCTION: Aid in depressing ribs

NERVE: Muscular branches of intercostals

ARTERY: Musculophrenic; posterior intercostals

REFERENCES:	GRAY	GRANT'S ATLAS
Muscle	410	410
Nerve	411, 979-982	406-409
Artery	610, 627	410

TRANSVERSUS THORACIS

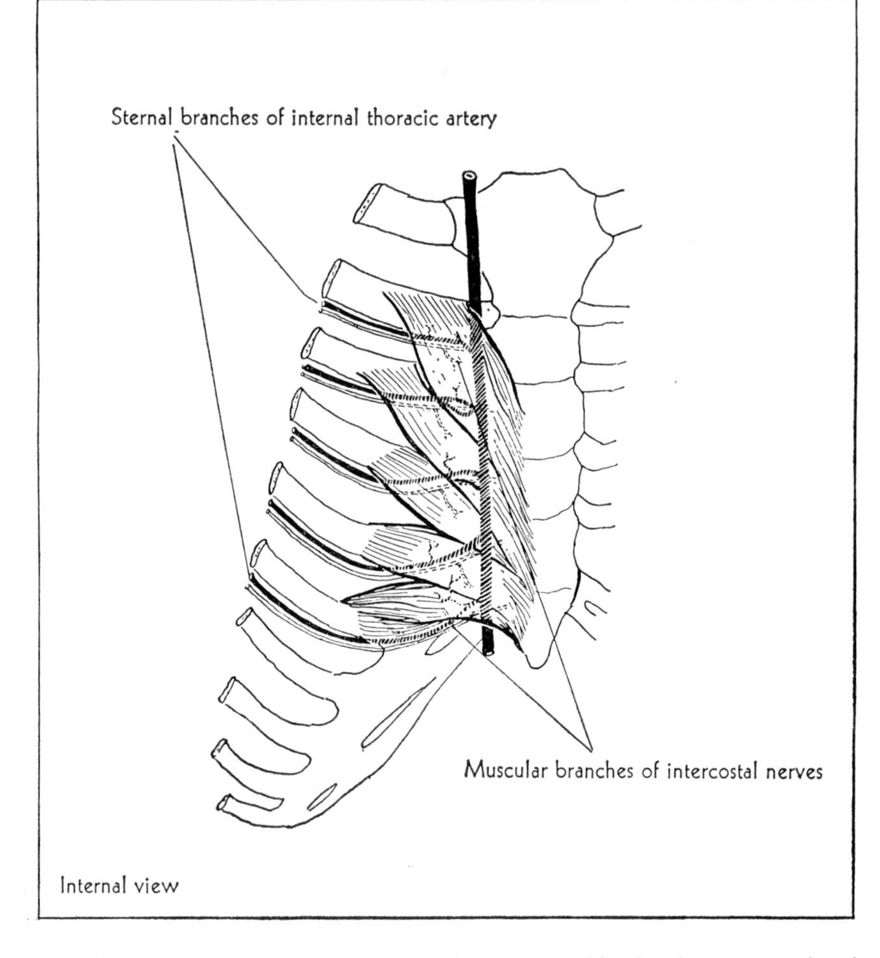

ORIGIN: Internal surface of xiphoid process and body of sternum to level of 3d costal cartilage and sternal ends of costal cartilages of lower 3 or 4 ribs

INSERTION: Lower borders and inner surfaces of 2d, 3d, 4th, 5th and 6th costal cartilages

FUNCTION: Depresses costal cartilages; muscle of expiration

NERVE: Anterior primary rami of upper 6 thoracic intercostal nerves

ARTERY: Sternal branches of internal thoracic, intercostals

REFERENCES:	GRAY	GRANT'S ATLAS
Muscle	411	406, 407
Nerve	411, 979-982	407
Artery	609, 627	407

LEVATORES COSTARUM

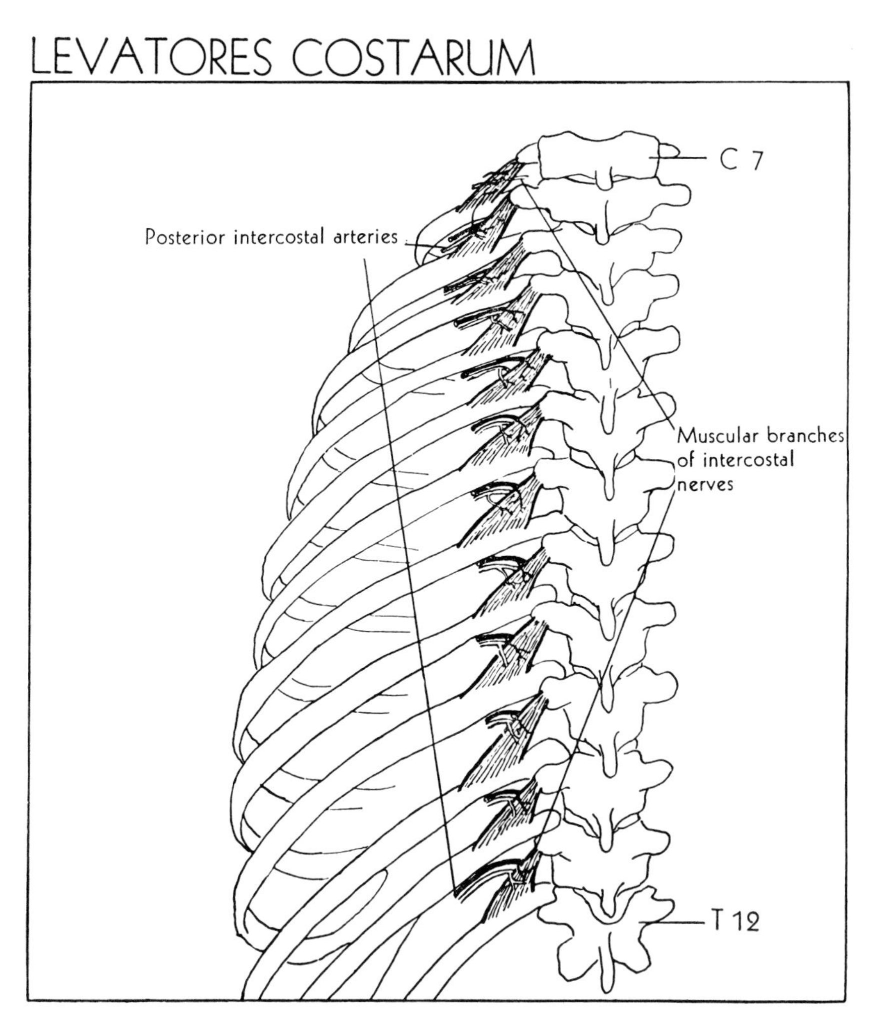

ORIGIN: 12 pairs, from transverse processes of 7th cervical and upper 11 thoracic vertebrae

INSERTION: Between tubercle and angle on outer surface of rib below vertebra of origin

FUNCTION: Rotators and lateral flexors of vertebral column, elevate ribs; assist in inspiration

NERVE: Anterior primary rami of thoracics

ARTERY: Posterior intercostals

REFERENCES:

	GRAY	GRANT'S ATLAS
Muscle	411	409, 411, 479
Nerve	411, 979-982	Not shown
Artery	627	Not shown

SERRATUS POSTERIOR SUPERIOR

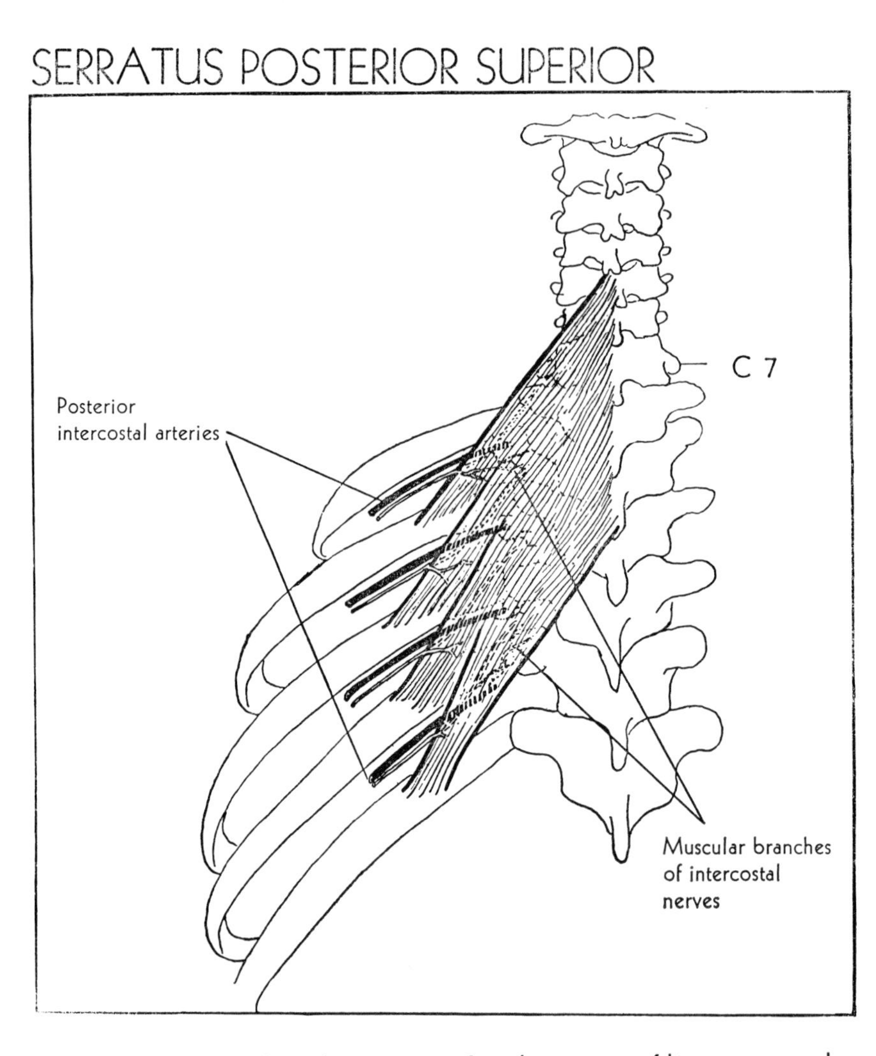

ORIGIN: By thin broad aponeurosis from lower part of ligamentum nuchae and spines of 7th cervical and upper 2 or 3 thoracic vertebrae

INSERTION: Upper border of 2d, 3d, 4th, and 5th ribs just beyond their angles

FUNCTION: Elevates ribs; muscle of inspiration

NERVE: Anterior primary rami of thoracics (upper 3)

ARTERY: Highest intercostal; posterior intercostals

REFERENCES:

	GRAY	GRANT'S ATLAS
Muscle	411	478
Nerve	411, 979-982	Not shown
Artery	610, 627	Not shown

SERRATUS POSTERIOR INFERIOR

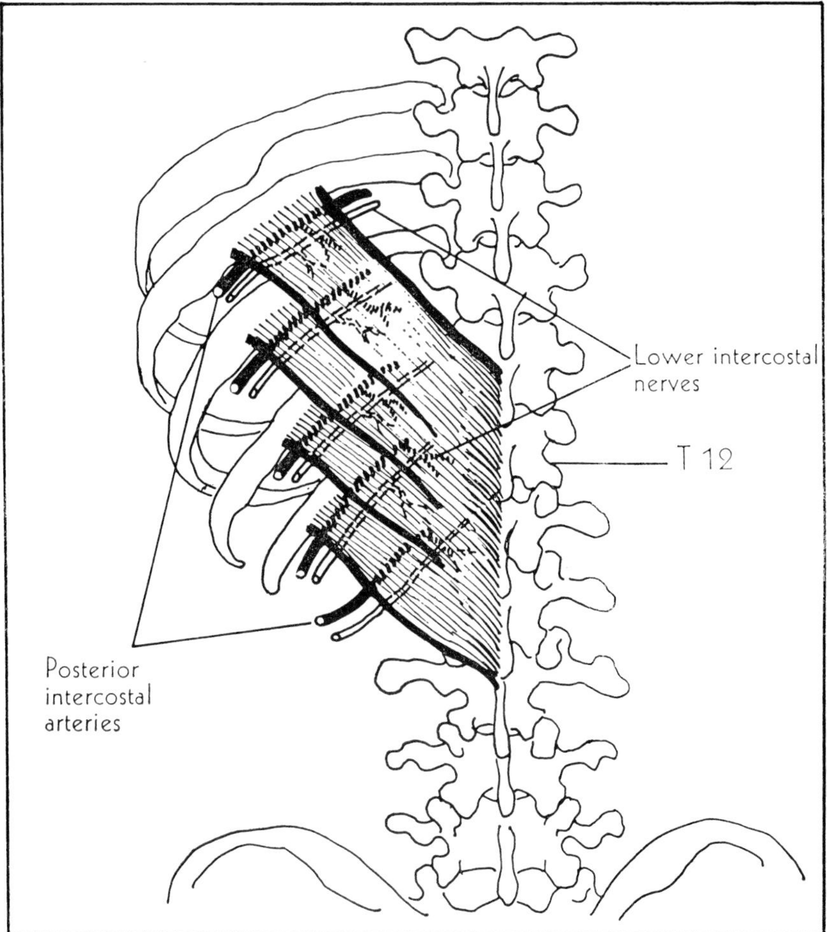

ORIGIN: Last 2 thoracic and first 2 lumbar spines, lumbar fascia
INSERTION: Inferior border of lower 4 ribs just beyond their angles
FUNCTION: Fixes lower ribs, draws them down and back; acts in expiration
NERVE: Anterior primary rami of lower 3 thoracic
ARTERY: Posterior intercostals

REFERENCES: GRAY		GRANT'S ATLAS
Muscle	412	173, 478
Nerve	412, 982	Not shown
Artery	627	Not shown

DIAPHRAGM

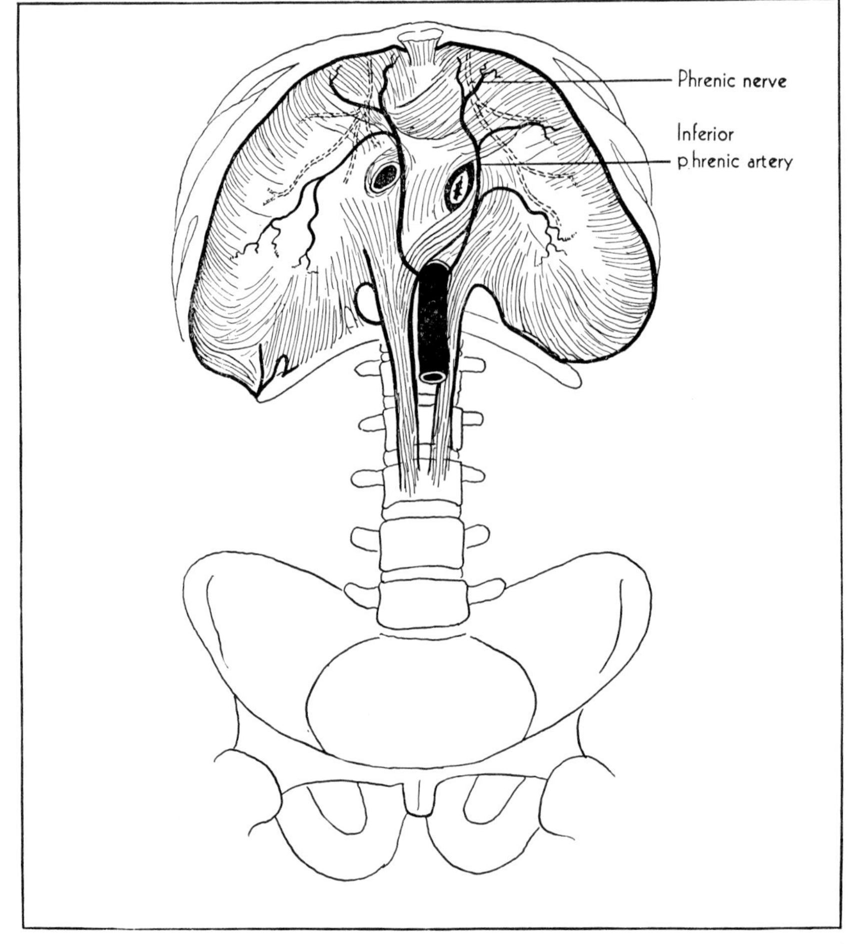

ORIGIN: Fleshy slips from internal surface of xiphoid process form sternal part; attachments to inner surfaces of cartilages and adjacent portions of lower 6 ribs on both sides form costal parts; aponeurotic lumbocostal arches and 2 crura from lumbar vertebrae form lumbar part

INSERTION: Muscle fibers arch upward and inward to end in tendinous fibers which form the central tendon

FUNCTION: Respiration; contractions increase capacity of thoracic basket

NERVE: Phrenic

ARTERY: Superior and inferior phrenic; branches of musculo-phrenic; pericardiaco-phrenic

REFERENCES:

	GRAY	GRANT'S ATLAS
Muscle	412	191, 407, 428, 431
Nerve	414, 955, 959, 963	428, 449, 474, 531
Artery	609, 610, 627, 640	179, 180, 428

OBLIQUUS EXTERNUS ABDOMINIS

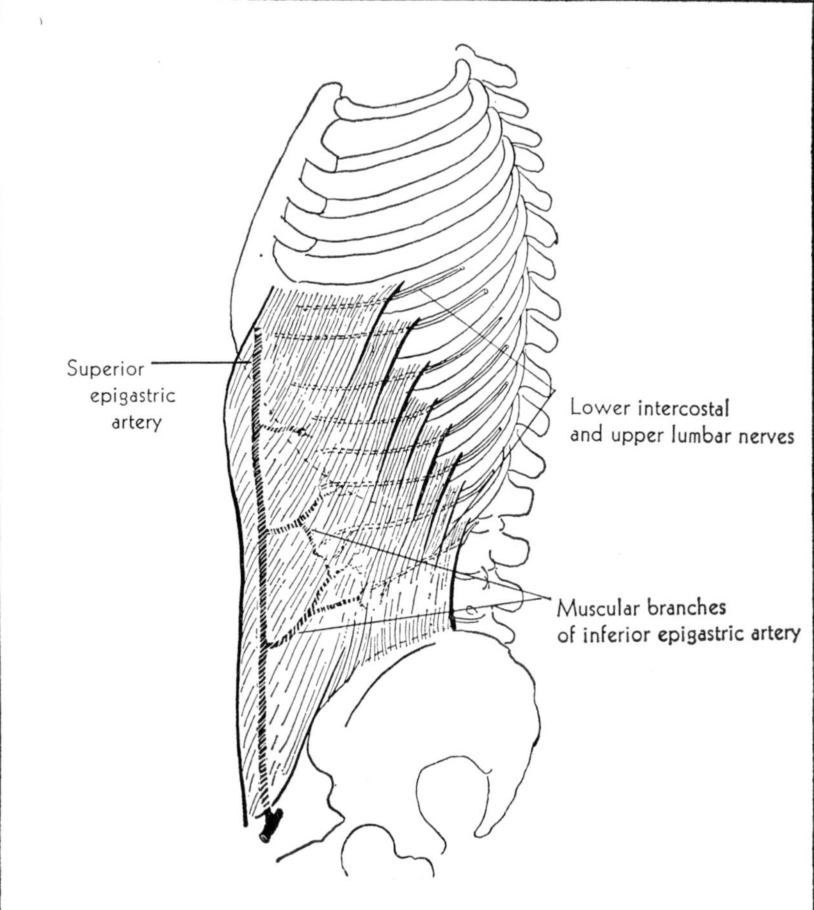

ORIGIN: By interdigitating slips from external surfaces of lower 8 ribs

INSERTION: Fleshy fibers from lowest ribs inserted into anterior half of outer lip of iliac crest, remainder inserted into wide aponeurosis of anterior abdominal wall

FUNCTION: Compresses abdomen, supports viscera, active in forced expiration

NERVE: Anterior primary rami of lower 6 thoracic and upper 2 lumbar

ARTERY: Muscular branches of superior and inferior epigastric

REFERENCES:	GRAY	GRANT'S ATLAS
Muscle	417	105, 173, 406
Nerve	418, 982-986	13, 107
Artery	610, 652	106, 205

OBLIQUUS INTERNUS ABDOMINIS

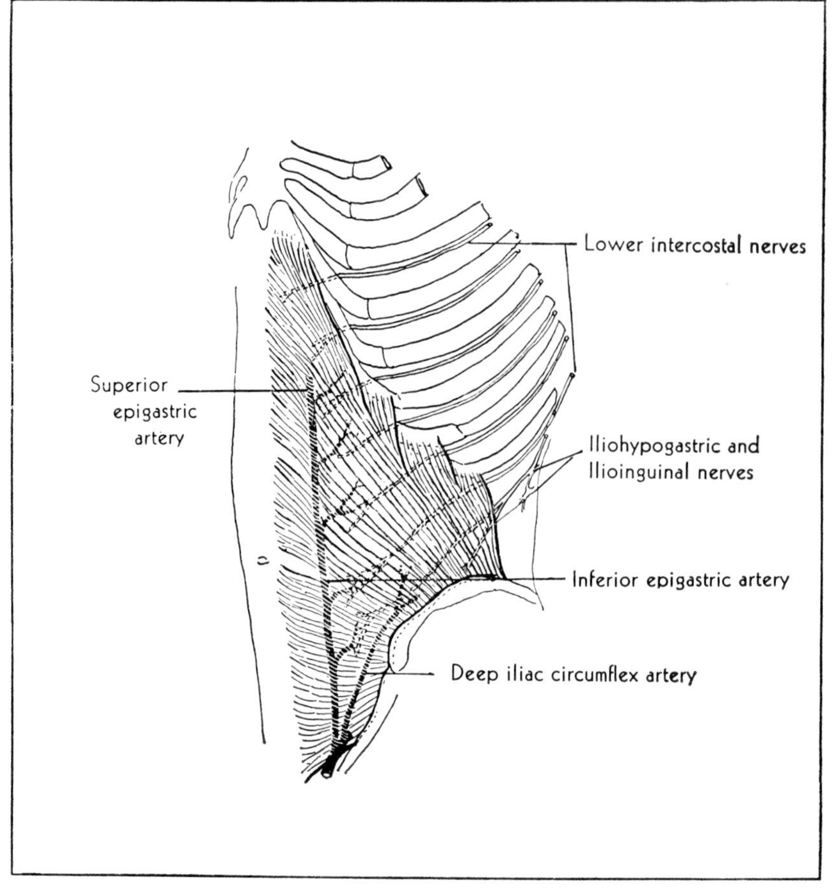

ORIGIN: Lumbar fascia, anterior two-thirds of middle lip of iliac crest, lateral two-thirds of inguinal ligament

INSERTION: Upper fibers into cartilages of last 3 ribs, remainder spread fan-like into an aponeurosis extending from 10th costal cartilage to pubic bone, forming linea alba in ventral mid-line

FUNCTION: Compresses abdomen, supports abdominal viscera, active in forced expiration

NERVE: Anterior primary rami of lower 6 thoracic and upper 2 lumbar, filaments from ilio-hypogastric and ilio-inguinal

ARTERY: Muscular branches of superior and inferior epigastric and deep circumflex iliac

REFERENCES: GRAY		GRANT'S ATLAS
Muscle	421	106, 110, 125
Nerve	423, 982-986	106, 110
Artery	610, 652	106, 205

CREMASTER

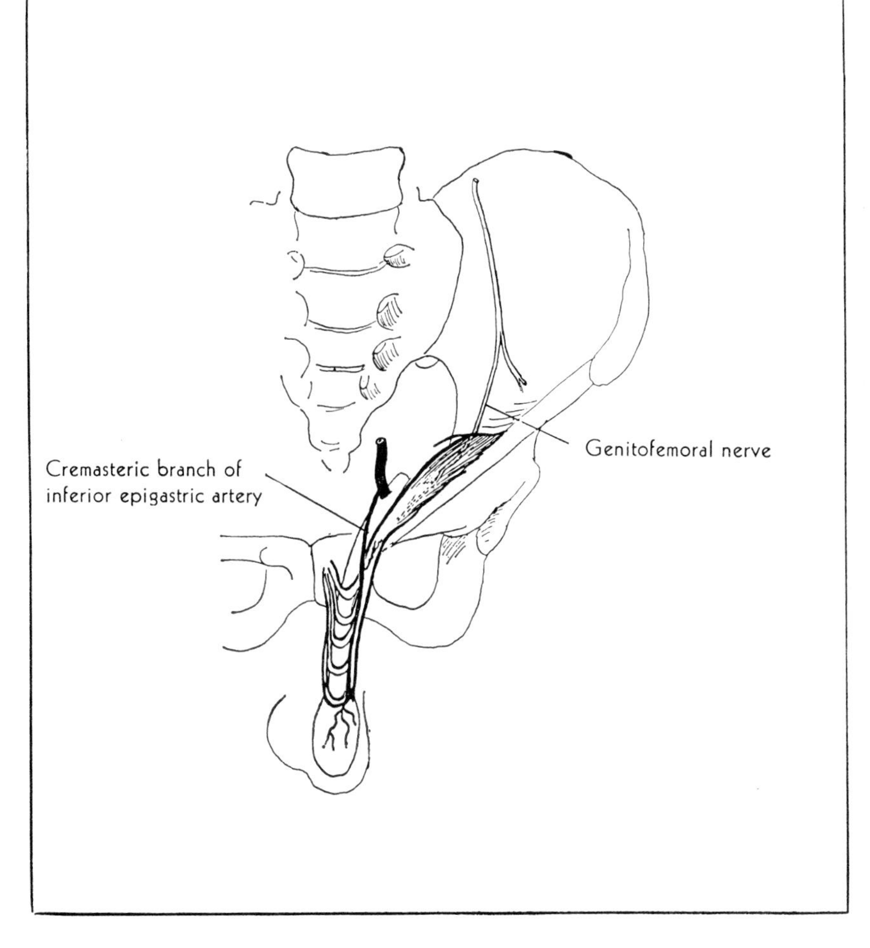

ORIGIN: Lower edge of internal oblique and middle of inguinal ligament
INSERTION: Pubic tubercle and crest of pubis
FUNCTION: Investment for testis and spermatic cord, retracts testis
NERVE: Genital branch of genito-femoral
ARTERY: Cremasteric branch of inferior epigastric

REFERENCES:	GRAY	GRANT'S ATLAS
Muscle	423	111, 117, 121
Nerve	423, 986	256
Artery	652	111, 112

TRANSVERSUS ABDOMINIS

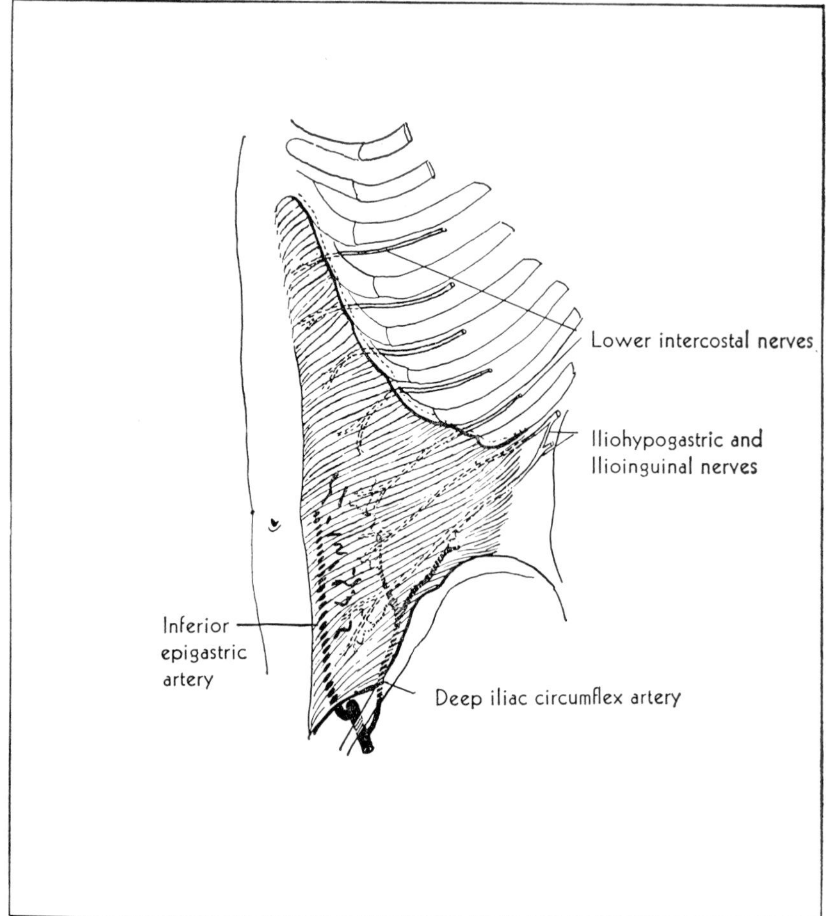

ORIGIN: Inner surfaces of costal cartilages of lower 6 ribs, middle layer of lumbar fascia, anterior two-thirds of inner lip of iliac crest and lateral third of inguinal ligament

INSERTION: By aponeurotic sheath with obliquii into linea alba

FUNCTION: Compresses abdomen, supports abdominal viscera, active in forced expiration

NERVE: Anterior primary rami of lower 6 intercostal, ilio-hypogastric and ilio-inguinal

ARTERY: Deep circumflex iliac, inferior epigastric

REFERENCES:	GRAY	GRANT'S ATLAS
Muscle	423	111, 174, 407
Nerve	424, 982-986	111, 174
Artery	652	111, 112

RECTUS ABDOMINIS

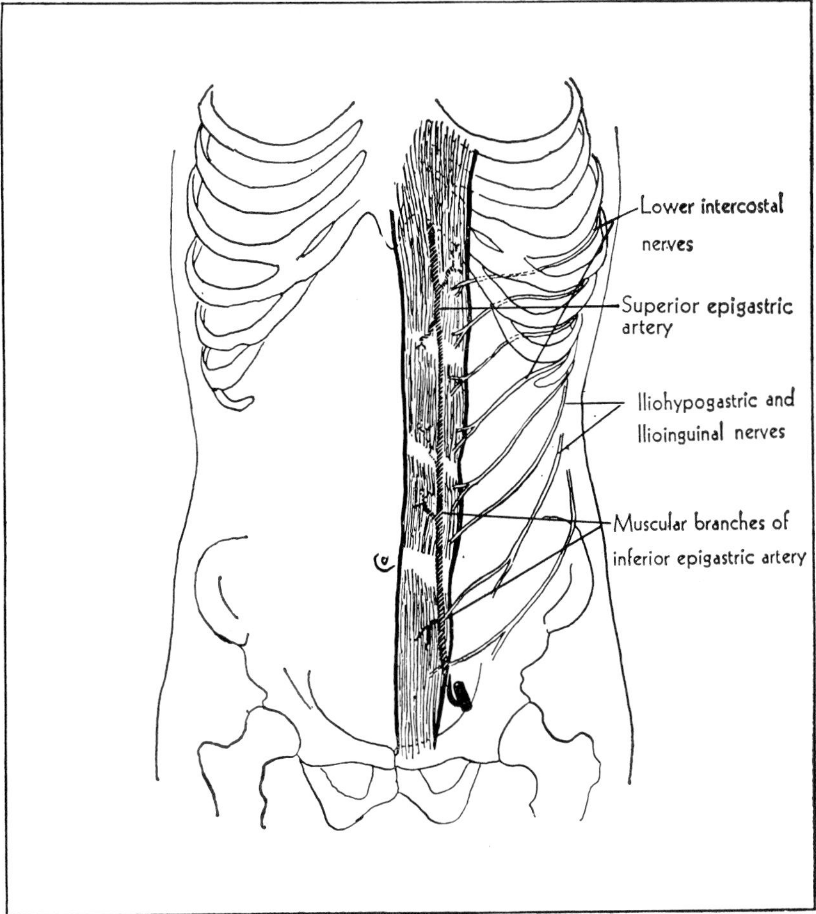

ORIGIN: Medial tendon from pubic symphysis, lateral tendon from crest of pubis

INSERTION: Anterior surface of xiphoid process and surface of costal cartilages of 5th, 6th, and 7th ribs

FUNCTION: Compresses abdomen, supports abdominal viscera, active in forced expiration, flexes pelvis and vertebral column

NERVE: Anterior primary rami of lower 6 intercostal, ilio hypogastric and ilio inguinal

ARTERY: Muscular branches of superior and inferior epigastric

REFERENCES:	GRAY	GRANT'S ATLAS
Muscle	424	105, 125, 406
Nerve	424, 982-986	106
Artery	610, 652	106

PYRAMIDALIS

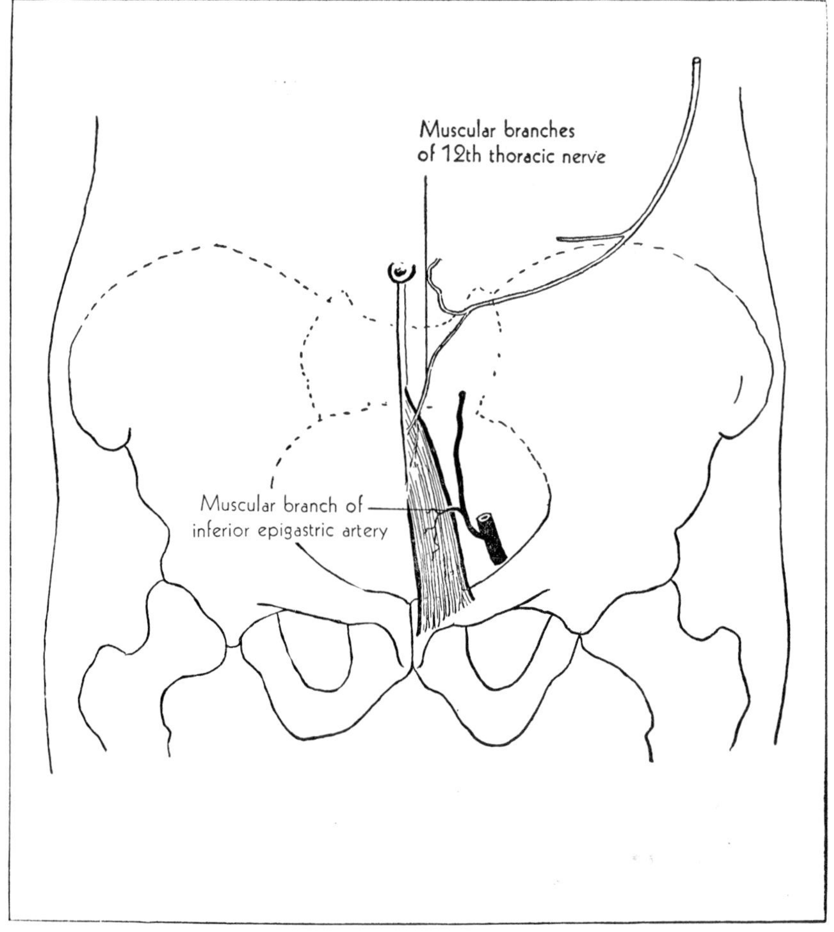

ORIGIN: Front of pubis and pubic ligament
INSERTION: On linea alba between umbilicus and pubis
FUNCTION: Compresses abdomen, supports abdominal viscera, active in forced expiration
NERVE: Muscular branches of subcostal (12th thoracic)
ARTERY: Muscular branches of inferior epigastric
REFERENCES: GRAY
Muscle 426
Nerve 427, 982
Artery 652

GRANT'S ATLAS
Not listed in 6th edition

QUADRATUS LUMBORUM

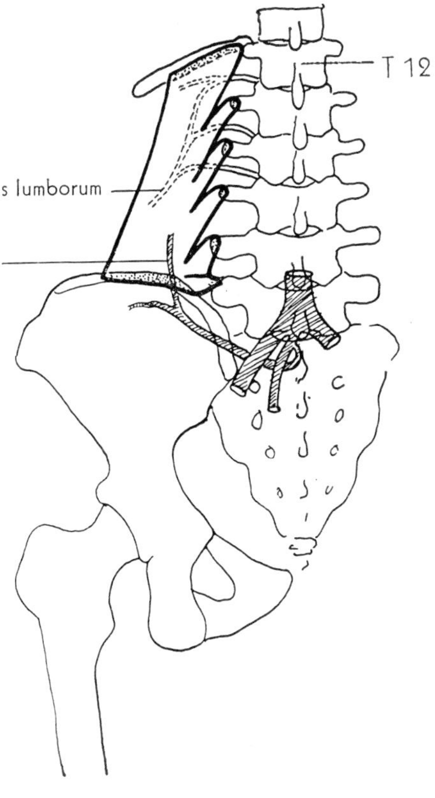

T 12

Nerves to quadratus lumborum

Lumbar branch of iliolumbar artery

ORIGIN: Iliolumbar ligament, posterior part of iliac crest

INSERTION: Lower border of last rib, transverse processes of upper 4 lumbar vertebrae

FUNCTION: Depresses last rib, aids in flexing trunk; acting singly it flexes column laterally, its action in breathing is questionable

NERVE: Subcostal, 1st, 2d, and 3d lumbar

ARTERY: Lumbar branch of iliolumbar

REFERENCES: GRAY	GRANT'S ATLAS
Muscle 431	175, 179, 190, 481
Nerve 431, 982-983	175, 190
Artery 649	Not shown

LEVATOR ANI

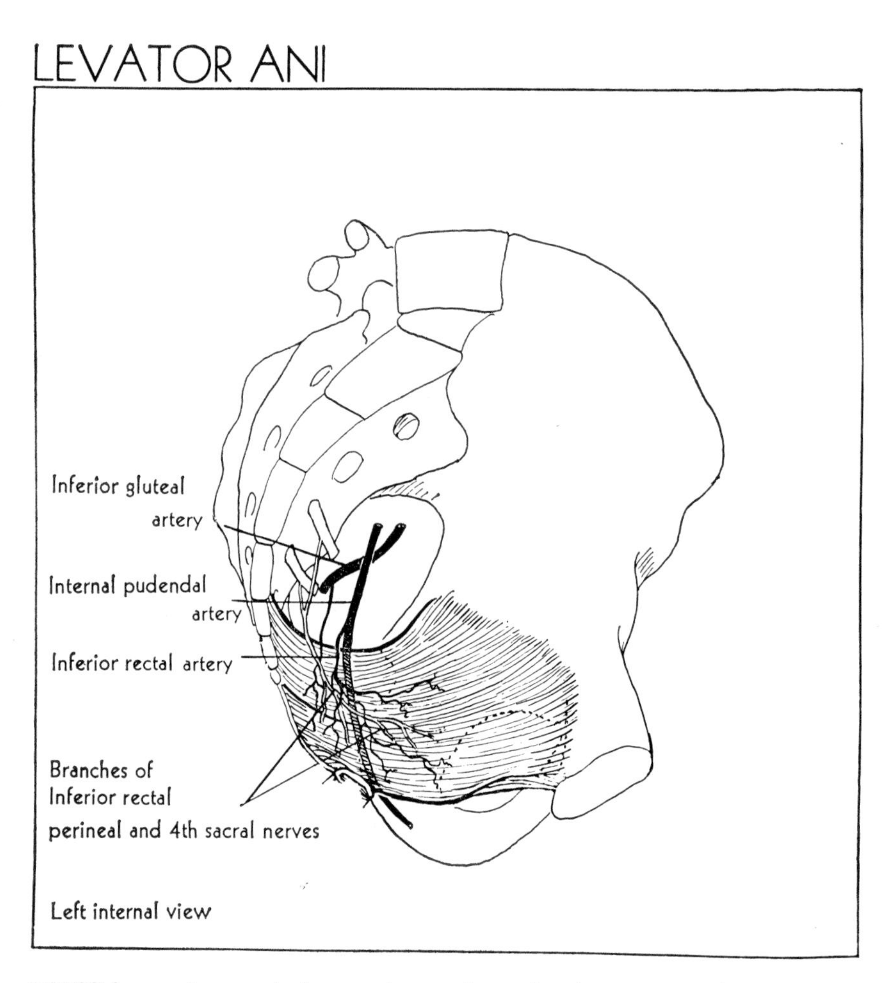

Left internal view

ORIGIN: Anteriorly from pelvic surface of pubis, posteriorly from inner surface of ischial spine, between these origins from obturator fascia

INSERTION: Front and sides of coccyx; more anteriorly fibers unite with those of opposite side in a median anococcygeal raphé extending between coccyx and anus, middle fibers into sides of rectum, anterior into perineal body

FUNCTION: Forms floor of pelvic cavity; constricts lower end of rectum and vagina, retains viscera in position; muscle of forced expiration

NERVE: Muscular branches of pudendal plexus

ARTERY: Muscular branches of internal pudendal, inferior rectal, inferior gluteal

REFERENCES: GRAY		GRANT'S ATLAS
Muscle	431	192, 205, 209, 231
Nerve	433, 993, 1003	204, 209, 213
Artery	648, 649	195

COCCYGEUS (Ischio-coccygeus)

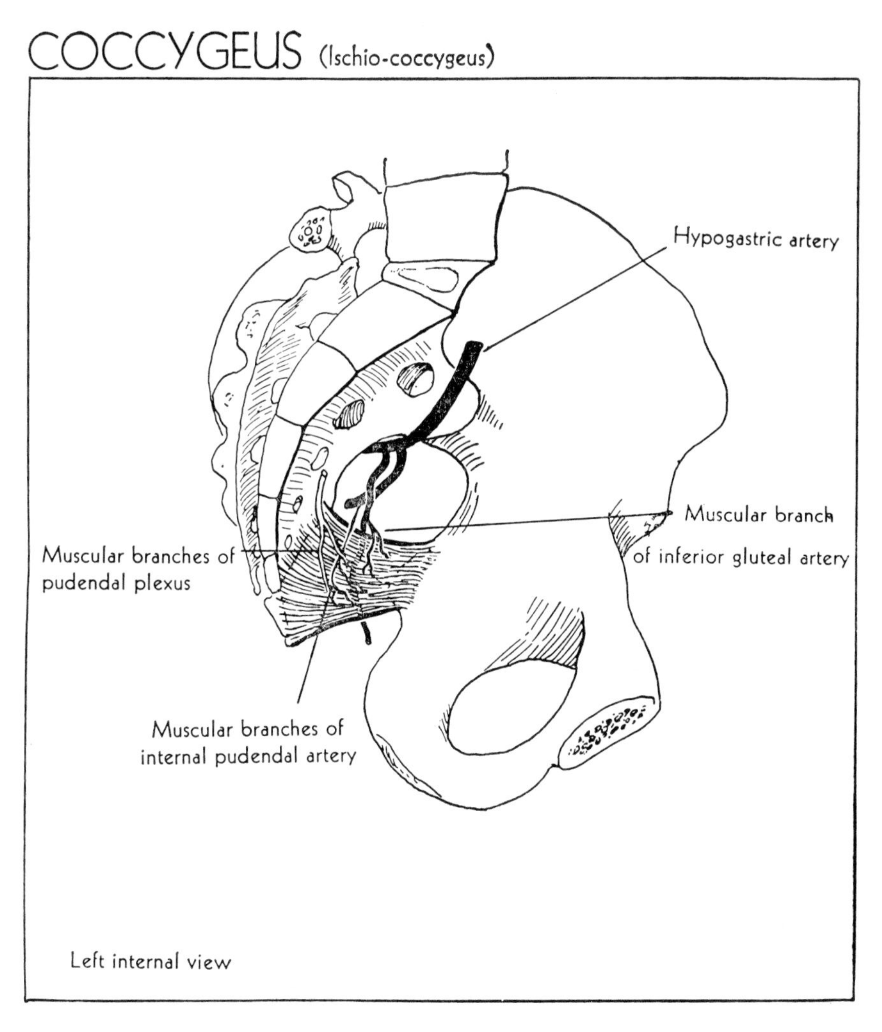

Left internal view

ORIGIN: Pelvic surface of ischial spine and sacrospinous ligament
INSERTION: Margin of coccyx and side of lowest segment of sacrum
FUNCTION: Pulls forward and supports coccyx; with levatores ani forms supporting muscular diaphragm for pelvic viscera
NERVE: Muscular branches of pudendal plexus
ARTERY: Muscular branch of internal pudendal, inferior gluteal

REFERENCES:	GRAY	GRANT'S ATLAS
Muscle	433	194
Nerve	433, 993, 1003	213
Artery	648, 649	Not shown

TRANSVERSUS PERINEI SUPERFICIALIS

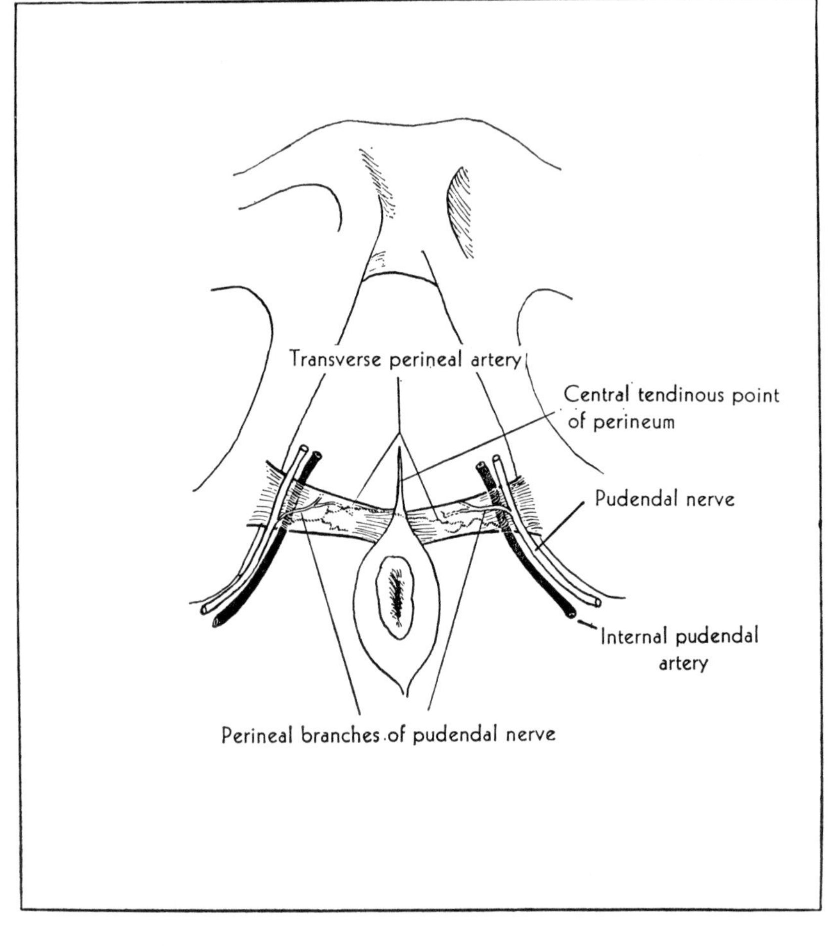

ORIGIN: Forepart of tuberosity of ischium
INSERTION: Central tendinous point of perineum; (perineal body)
FUNCTION: Fix central tendinous point of perineum
NERVE: Perineal branches of pudendal
ARTERY: Perineal branches of internal pudendal

REFERENCES:	GRAY	GRANT'S ATLAS
Muscle	441, 444	192, 227
Nerve	443, 444, 993, 1003	192
Artery	648	192

BULBOCAVERNOSUS (male)

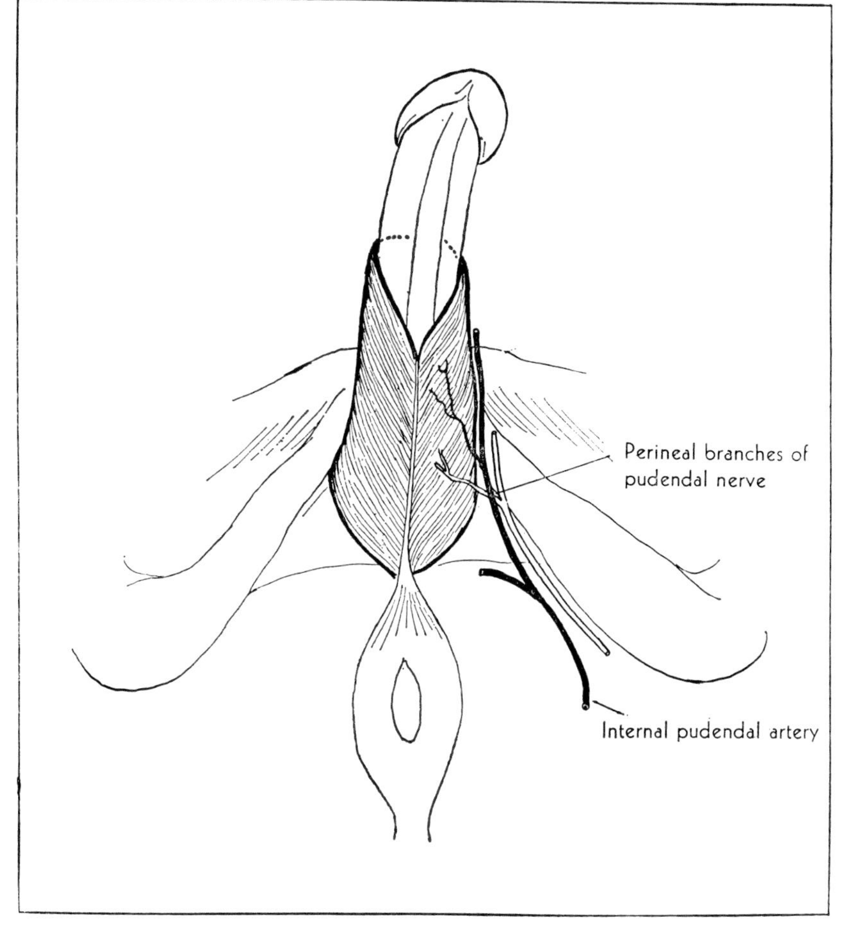

ORIGIN: Central tendinous point of perineum and median raphé in front
INSERTION: Lower surface of perineal membrane, dorsal surface of corpus spongiosum, deep fascia on dorsum of penis
FUNCTION: Accessory muscle of penile erection, empties urethra at end of micturition
NERVE: Perineal branches of pudendal
ARTERY: Perineal branches of internal pudendal

REFERENCES:	GRAY	GRANT'S ATLAS
Muscle	441	192, 203, 204
Nerve	443, 993, 1003	205
Artery	648	192

BULBOCAVERNOSUS (female)

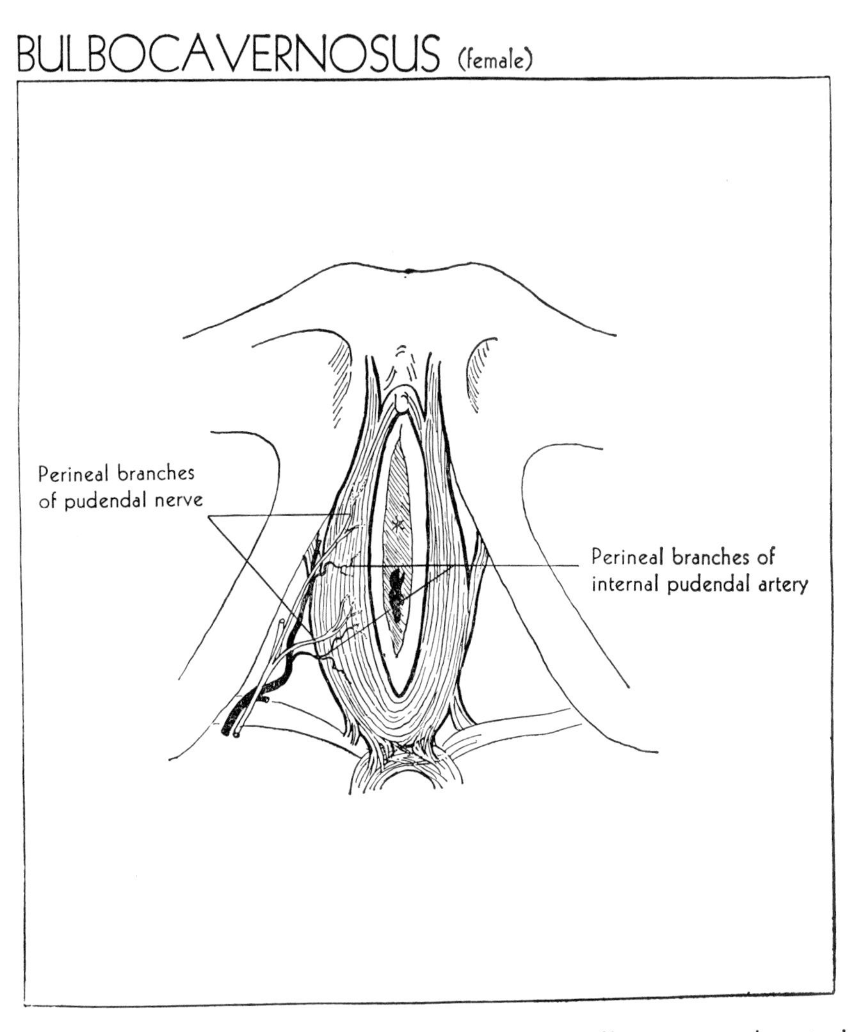

ORIGIN: Central tendinous point of perineum; fibers surround vaginal orifice and bulb of vestibule

INSERTION: Sides of pubic arch and corpora cavernosa clitoridis

FUNCTION: Constricts orifice of vagina and aids in erection of clitoris

NERVE: Perineal branches of pudendal

ARTERY: Perineal branches of internal pudendal

REFERENCES:

	GRAY	GRANT'S ATLAS
Muscle	444	227, 228
Nerve	444, 993, 1003	Not shown
Artery	648, 649	227

ISCHIOCAVERNOSUS

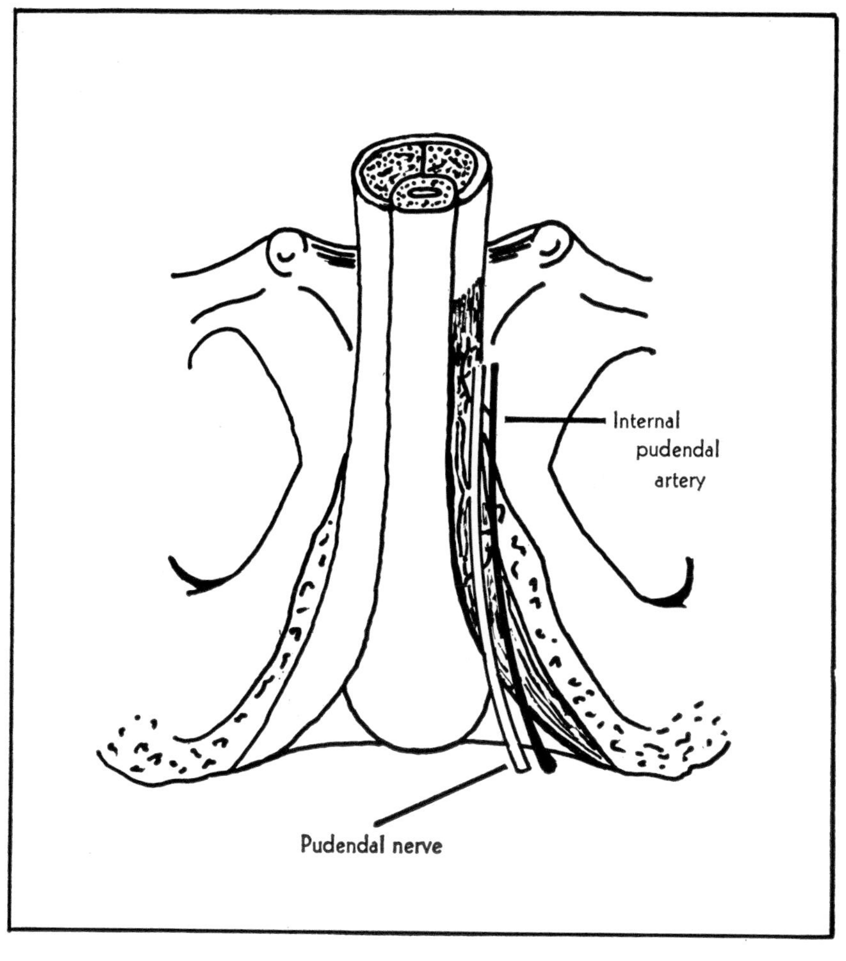

ORIGIN: Medial side of ischial tuberosity, rami of pubis and ischium on either side of crus of penis or clitoris

INSERTION: On each side of crus penis or clitoris, into corpus cavernosum and margin of pubic arch

FUNCTION: Impedes venous return by compressing crus penis, chief agent in erection of penis or clitoris

NERVE: Perineal branches of pudendal

ARTERY: Perineal branches of internal pudendal

REFERENCES:

	GRAY	GRANT'S ATLAS
Muscle	442	192, 227
Nerve	443, 444, 993, 1003	Not shown
Artery	648	192

TRANSVERSUS PERINEI PROFUNDUS (Male)*

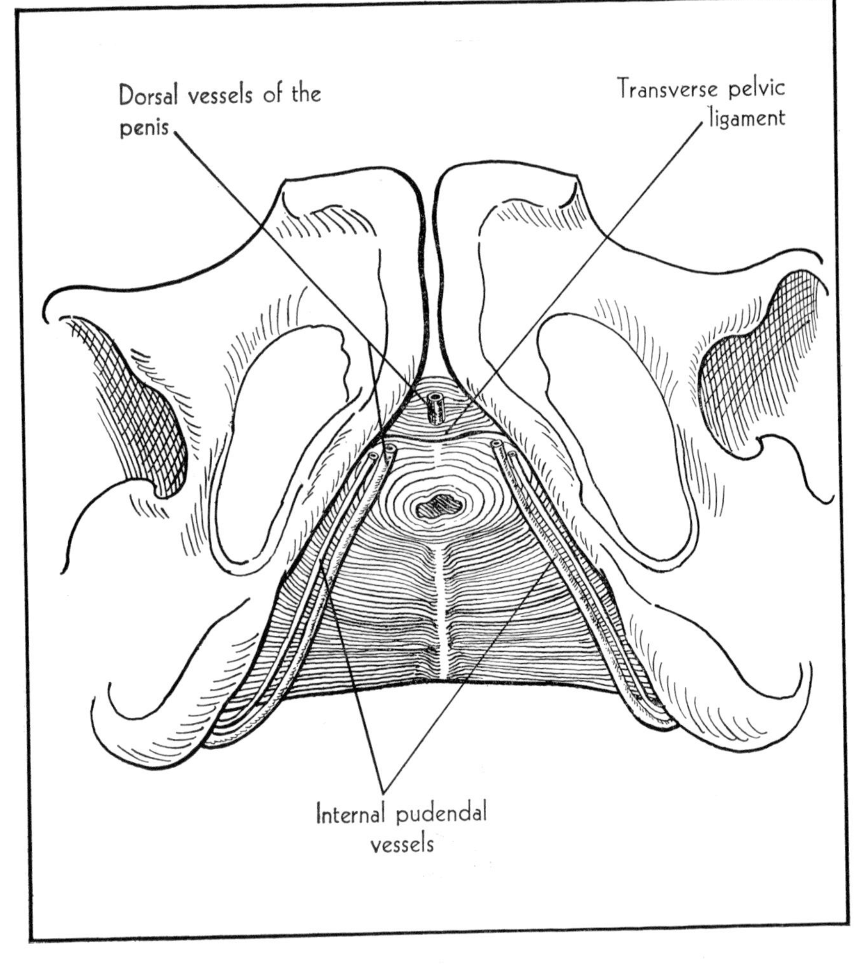

ORIGIN: From the inner surface of the inferior ischial rami
INSERTION: A medial tendinous raphé and perineal body
FUNCTION: Fixation of perineal body; support of prostate; augments sphincter action
NERVE: Perineal branches of the pudendal
ARTERY: Perineal branches of the internal pudendal

REFERENCES: GRAY	GRANT'S ATLAS
Muscle 443	201
Nerve 443, 993, 1003	216
Artery 648	216

***The Sphincter Urethrae lies in the same plane. (Page 88)**

TRANSVERSUS PERINEI PROFUNDUS (Female)*

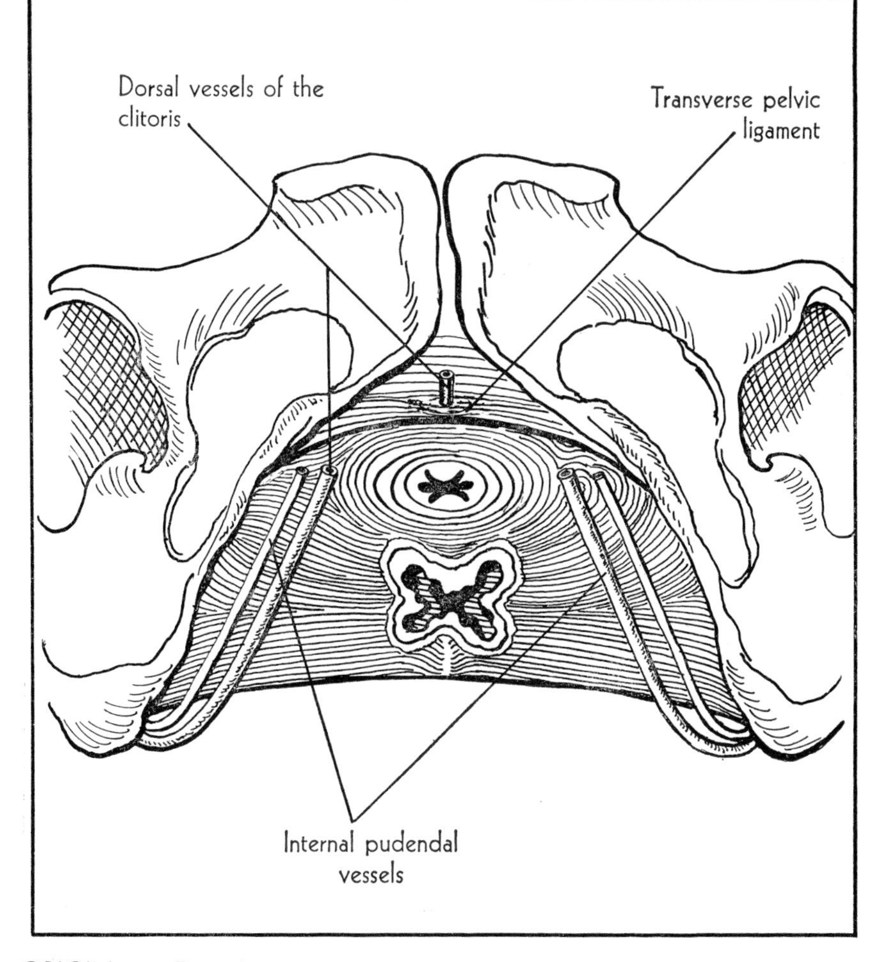

ORIGIN: From the inner surface of the inferior ischial rami
INSERTION: Sides of the vagina
FUNCTION: Assists in fixation of perineal body; augments sphincter action; supports vagina
NERVE: Perineal branches of the pudendal
ARTERY: Perineal branches of the internal pudendal

REFERENCES:	GRAY	GRANT'S ATLAS
Muscle	444	228, 229
Nerve	444, 993, 1003	Not shown
Artery	648, 649	Not shown

*The Sphincter Urethrae lies in the same plane. (Page 88)

SPHINCTER URETHRAE

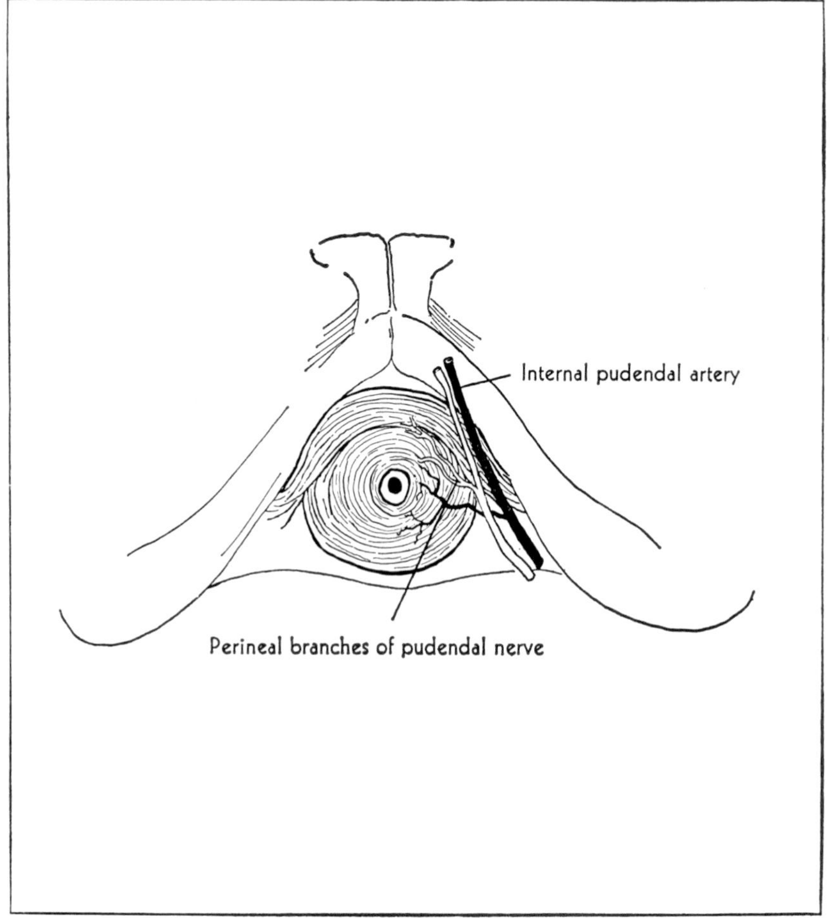

ORIGIN: External fibers from junction of inferior pubic and ischial rami and adjacent fascia, internal fibers pass medially to surround membranous urethra

INSERTION: (Male), into median raphé in front and behind urethra; (female), encloses urethra, attaches also to sides of vagina

FUNCTION: Compresses urethra at end of micturition

NERVE: Perineal branches of pudendal

ARTERY: Perineal branch of internal pudendal

REFERENCES:	GRAY	GRANT'S ATLAS
Muscle	443, 444	203, 230
Nerve	443, 444, 993, 1003	Not shown
Artery	648	Not shown

SPHINCTER ANI EXTERNUS*

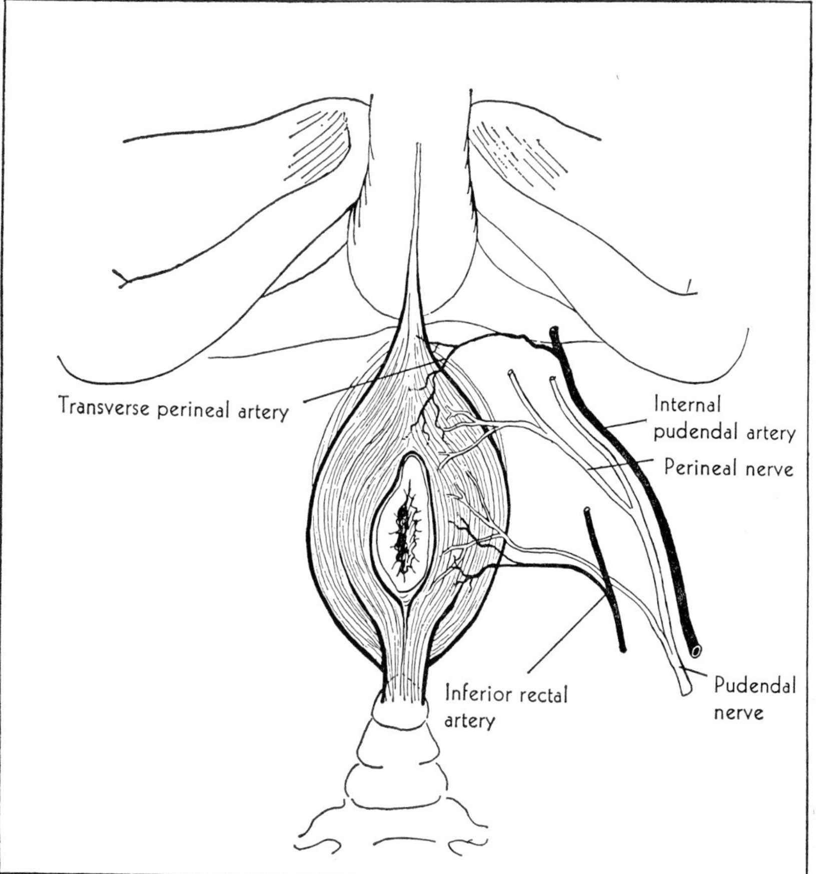

ORIGIN and INSERTION: Superficial fibers from anococcygeal raphé; fibers decussate around anus and meet anteriorly in central point of perineum
Deeper fibers surround anal canal, attach posteriorly to coccyx and anteriorly to central point of perineum; may decussate in front of anus and join superficial transversus perinaei muscle

FUNCTION: Closes anal orifice, contraction increased voluntarily

NERVE: Perineal and inferior rectal branches of pudendal

ARTERY: Inferior rectal and transverse perineal

REFERENCES:	GRAY	GRANT'S ATLAS
Muscle	445	192, 196, 203
Nerve	445, 993, 1003	192
Artery	648	192

*The sphincter ani internus is a ring of involuntary muscle which surrounds the terminal portion of the anal canal. Its fibers are continuous with the circular fibers of the intestine. The corrugator cutis ani comprises a group of involuntary muscle fibers radiating subcutaneously from the anal orifice

LEVATOR PALPEBRAE SUPERIORIS

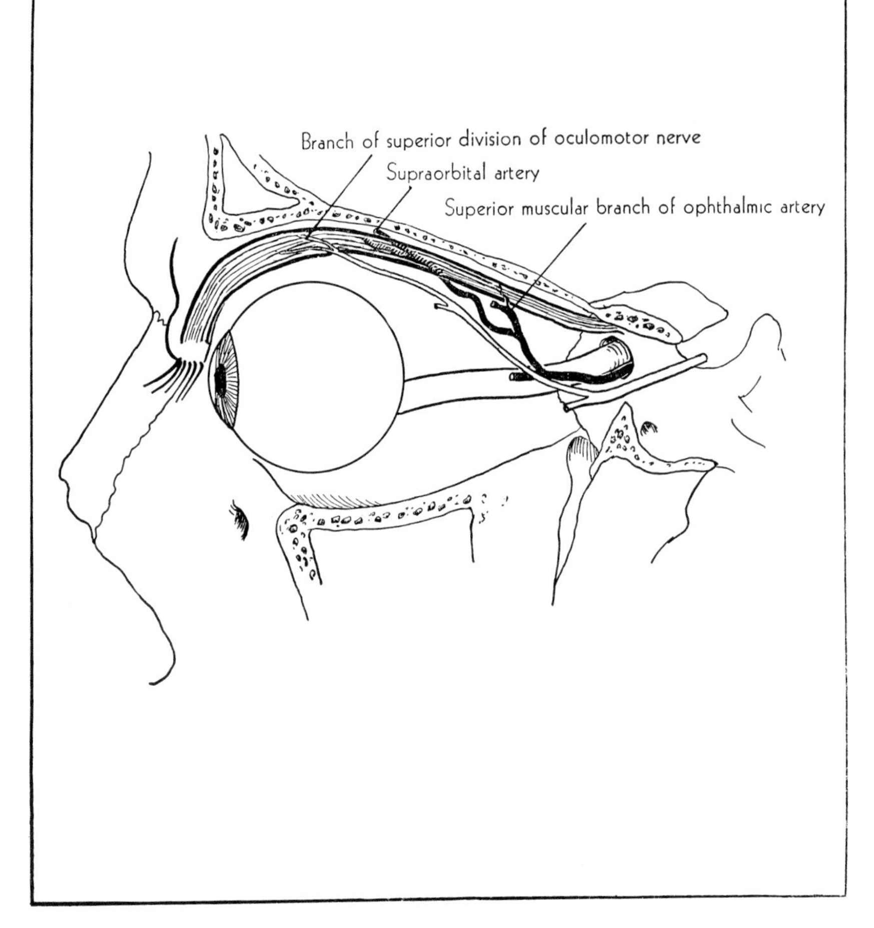

ORIGIN: Roof of orbit in front of optic foramen

INSERTION: Ends in aponeurotic sheath which splits into 3 lamellae:

1) superficial lamella to deep surface of upper eyelid
2) middle lamella (smooth muscle) to upper margin of superior tarsus
3) deep lamella with expansion of sheath of superior rectus to superior fornix of conjunctiva

FUNCTION: Elevates upper eyelid voluntarily, middle lamella acts involuntarily

NERVE: Superior division of oculomotor

ARTERY: Superior muscular branch of ophthalmic division of internal carotid; supraorbital

REFERENCES:	GRAY	GRANT'S ATLAS
Muscle	1059	470, 517, 520
Nerve	1062, 910	525, 653
Artery	593, 596	520

RECTI: SUPERIOR AND INFERIOR

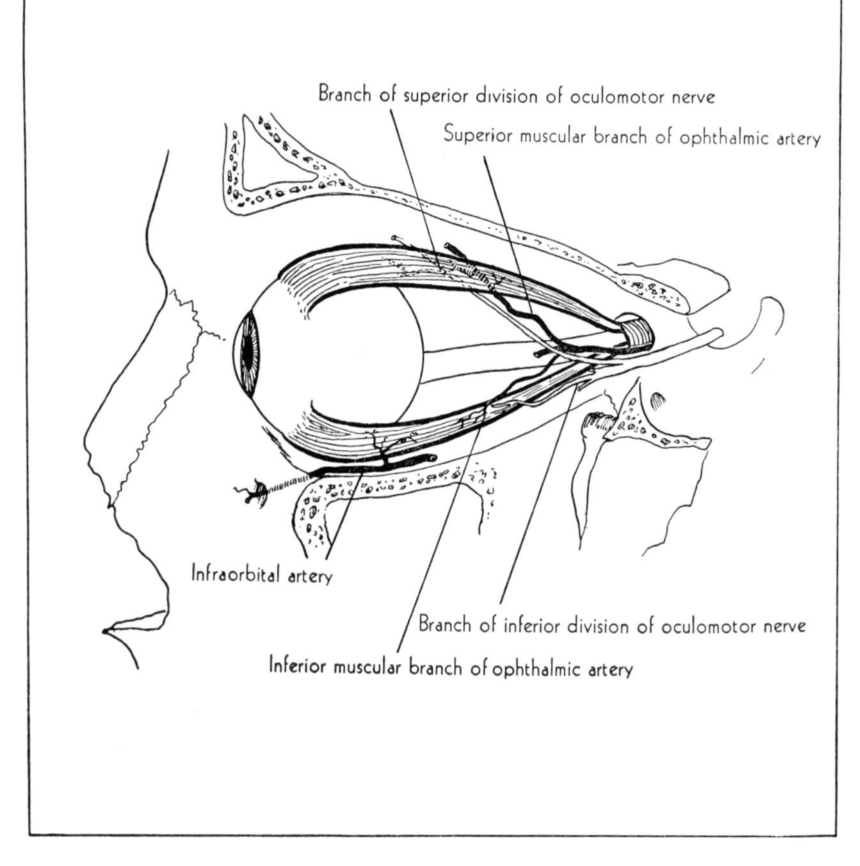

ORIGIN: From fibrous ring, the annulus tendineus communis, which surrounds the upper, medial and lower margins of the optic foramen

INSERTION: Into sclera about 6mm. from margin of cornea, in a vertical plane just medial to the vertical axis of the eyeball

FUNCTION: Superior rectus elevates, adducts and medially rotates the eyeball; inferior rectus depresses, adducts and laterally rotates the eyeball

NERVE: Rectus superior, branch of superior division of oculomotor; rectus inferior, branch of inferior division of oculomotor

ARTERY: Superior rectus: supraorbital and superior muscular branch of ophthalmic; inferior rectus: infraorbital branch of maxillary, inferior muscular branch of ophthalmic

REFERENCES:	GRAY	GRANT'S ATLAS
Muscle	1060	517-523
Nerve	1062, 910	525, 653
Artery	590, 593, 596	Not shown

RECTI: LATERALIS AND MEDIALIS

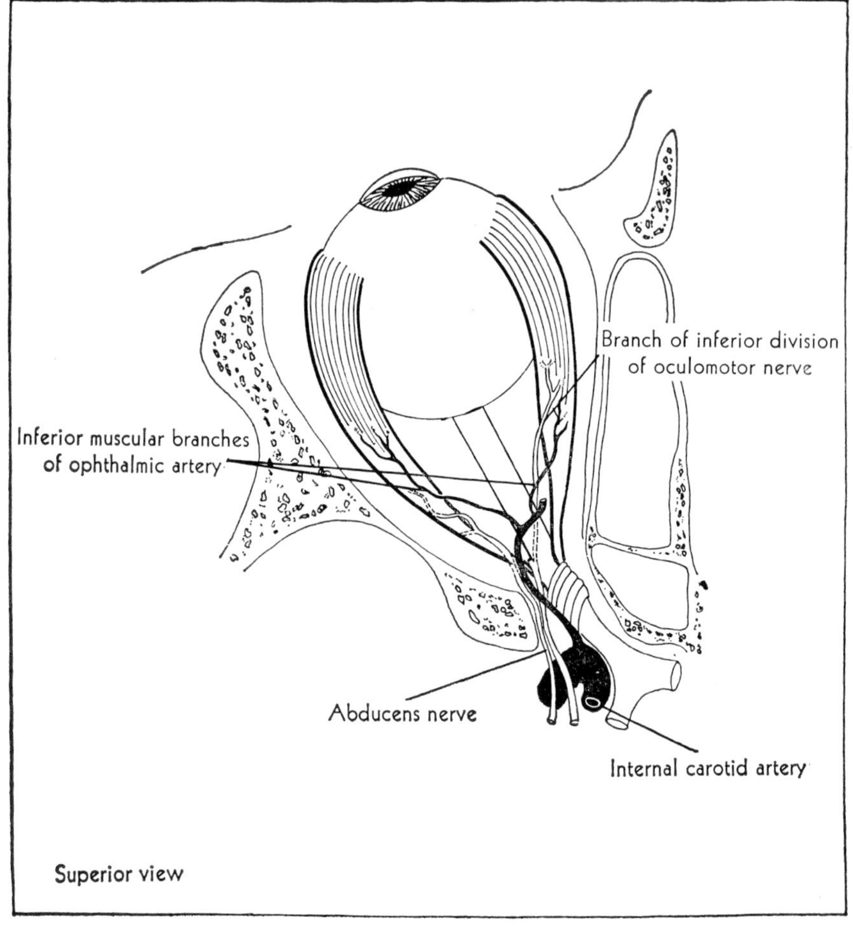

ORIGIN: From fibrous ring, the annulus tendineus communis, which surrounds the upper, medial and lower margins of the optic foramen

INSERTION: Laterally and medially into sclera about 6mm. from margin of cornea, in the transverse plane of the eyeball

FUNCTION: Move the eyes to right or left in a horizontal plane

NERVE: Lateralis by abducens; medialis by oculomotor (inferior division)

ARTERY: Inferior muscular branch of ophthalmic

REFERENCES:	GRAY	GRANT'S ATLAS
Muscle	1060	517-523
Nerve	1062, 910, 924	521, 525, 653
Artery	596	Not shown

OBLIQUUS SUPERIOR

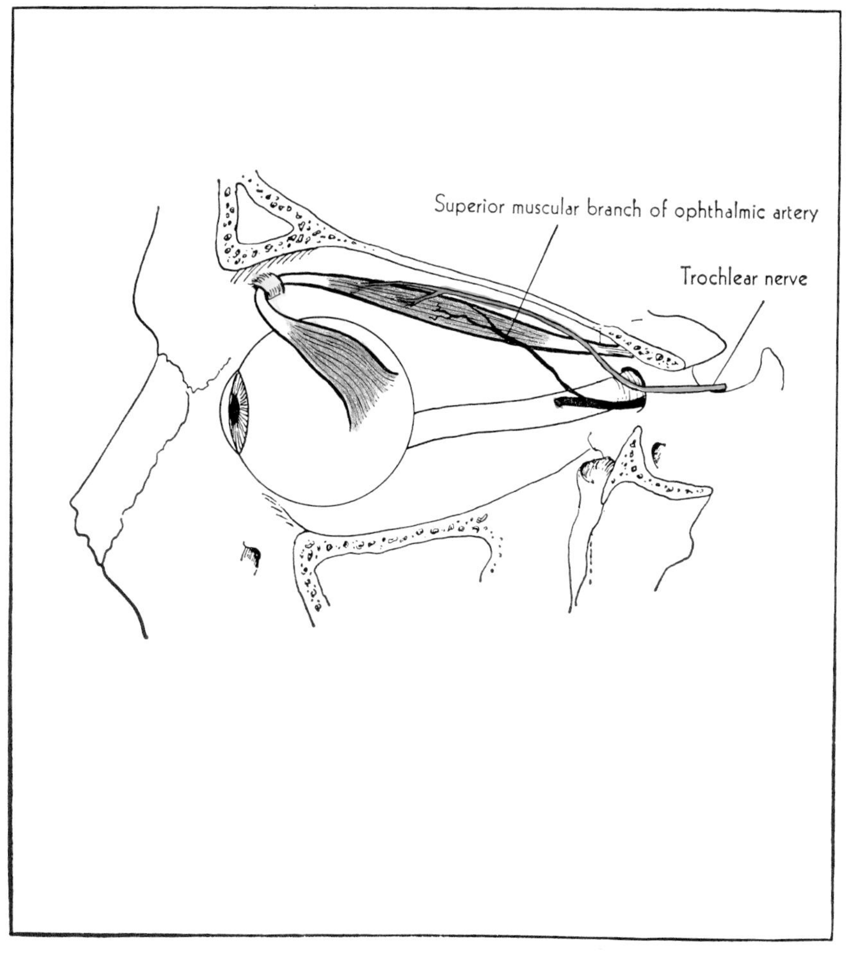

ORIGIN: Above margin of optic foramen above and medial to origin of rectus superior. At anterior part of orbit its tendon passes through a fibricartilaginous pulley, muscle then passes laterally and backward

INSERTION: Into sclera between superior and lateral recti

FUNCTION: Aids in rotating eyeball downward and laterally

NERVE: Trochlear

ARTERY: Superior muscular branches of ophthalmic

REFERENCES:

	GRAY	GRANT'S ATLAS
Muscle	1060	519-522.1, 525
Nerve	1062, 911-912	521, 525, 653
Artery	596	520

OBLIQUUS INFERIOR

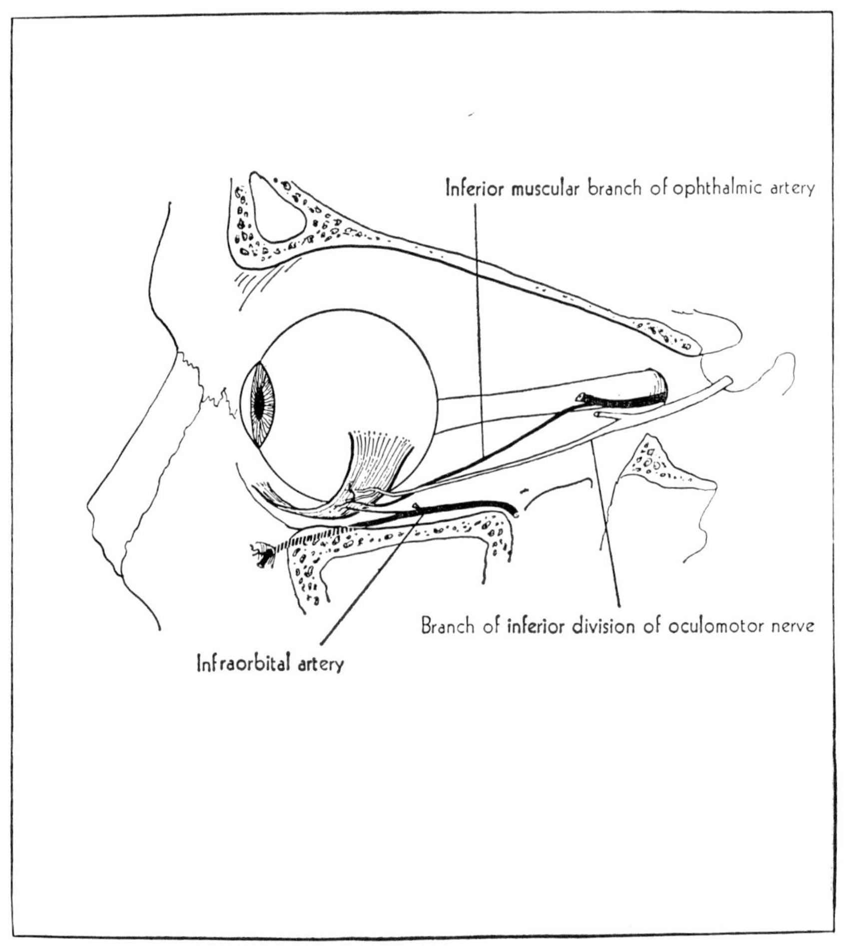

ORIGIN: Floor of orbit lateral to naso-lacrimal canal
INSERTION: Into sclera between superior and lateral recti further back than superior oblique
FUNCTION: Aids in rotating eyeball upward and laterally
NERVE: Inferior division of oculomotor
ARTERY: Infraorbital branch of maxillary, inferior muscular branch of ophthalmic

REFERENCES: GRAY	GRANT'S ATLAS
Muscle 1060	519, 522.1, 525
Nerve 1062, 910	521, 525, 653
Artery 590, 596	Not shown

TENSOR TYMPANI

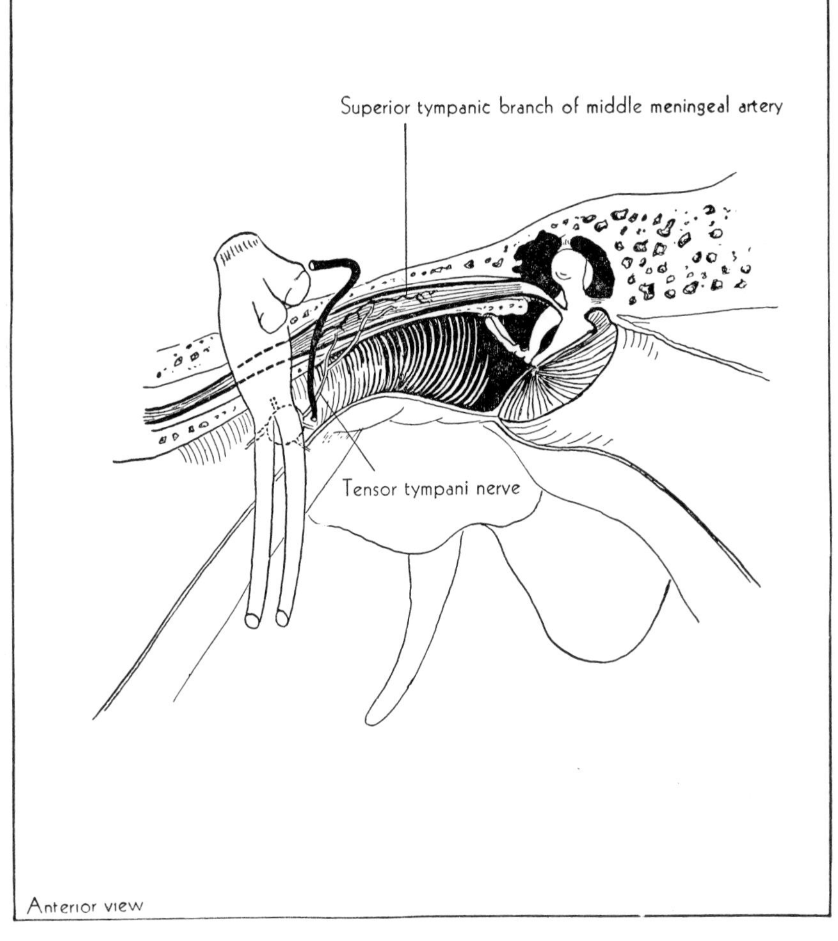

ORIGIN: The muscle lies in the bony canal above the osseus portion of the auditory tube. Arises from cartilage of auditory tube, adjoining part of greater wing of sphenoid and osseus canal in which it lies

INSERTION: By a slender tendon which crosses the tympanic cavity and inserts near the root of the handle of the malleus

FUNCTION: Tenses tympanic membrane by drawing it medially

NERVE: Branch of mandibular division of trigeminal through otic ganglion*

ARTERY: Superior tympanic branch of middle meningeal division of maxillary

REFERENCES: GRAY	GRANT'S ATLAS
Muscle 1083	641, 641.2
Nerve 1084, 920, 934	654
Artery 590	Not shown

*Contiguous relationship only!

STAPEDIUS

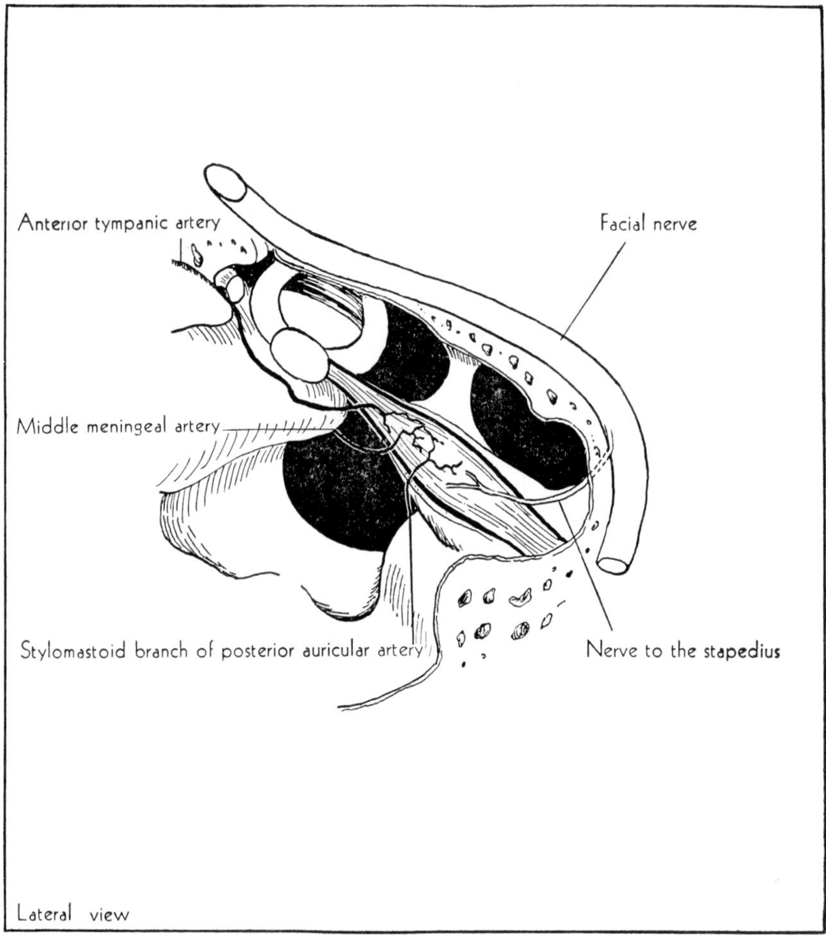

ORIGIN: Walls of cavity of pyramid; tendon emerges from apex of pyramid

INSERTION: Posterior surface of neck of stapes

FUNCTION: Draws back head of stapes and causes tilting of anterior part of base of stapes toward tympanic cavity; probably increases tension of fluid within inner ear.

NERVE: Nerve to the stapedius, which is a branch of the facial within the facial canal

ARTERY: Stylomastoid branch of posterior auricular, anterior tympanic and middle meningeal branches of maxillary

REFERENCES:	GRAY	GRANT'S ATLAS
Muscle	1083	636
Nerve	1084, 925, 928	657
Artery	585, 589	Not shown

CRICOTHYREOIDEUS

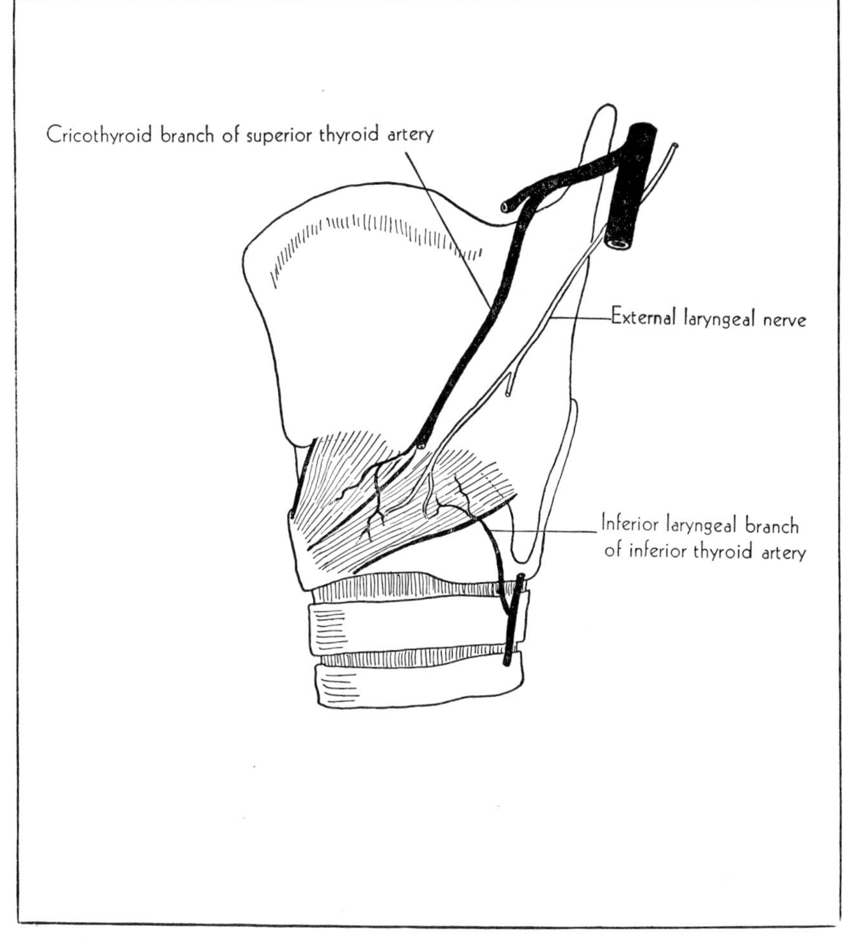

ORIGIN: Lateral surface of cricoid cartilage; fibers diverge into two groups

INSERTION: Lower group, oblique fibers slant backward to anterior margin of inferior horn of thyroid cartilage; upper, more erect fibers into lower border of lamina of thyroid cartilage

FUNCTION: Elevates arch of cricoid cartilage and tilts back its upper border, thereby tensing and elongating the vocal cords

NERVE: External laryngeal branch of superior laryngeal

ARTERY: Cricothyroid branch of superior thyroid; inferior laryngeal branch of inferior thyroid

REFERENCES: GRAY	GRANT'S ATLAS
Muscle 1128	530, 621, 628
Nerve 1131, 941	530, 622
Artery 581, 605	529

CRICOARYTENOIDEUS POSTERIOR

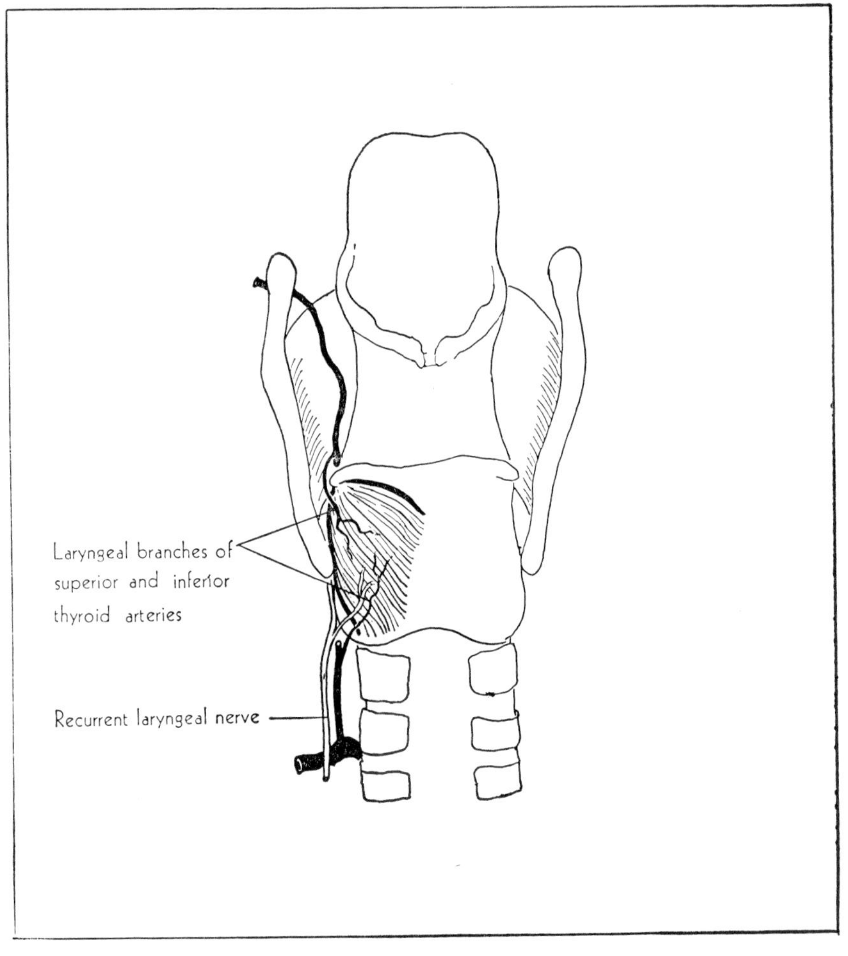

ORIGIN: Broad depression on posterior surface of lamina of cricoid cartilage

INSERTION: Fibers converge upon muscular process of arytenoid cartilage

FUNCTION: Opens glottis by separating the vocal folds by rotation and tilting of arytenoid cartilages

NERVE: Recurrent laryngeal

ARTERY: Laryngeal branches of superior and inferior thyroid

REFERENCES:	GRAY	GRANT'S ATLAS
Muscle	1128	627
Nerve	1131, 941	627
Artery	581, 605	Not shown

CRICOARYTENOIDEUS LATERALIS

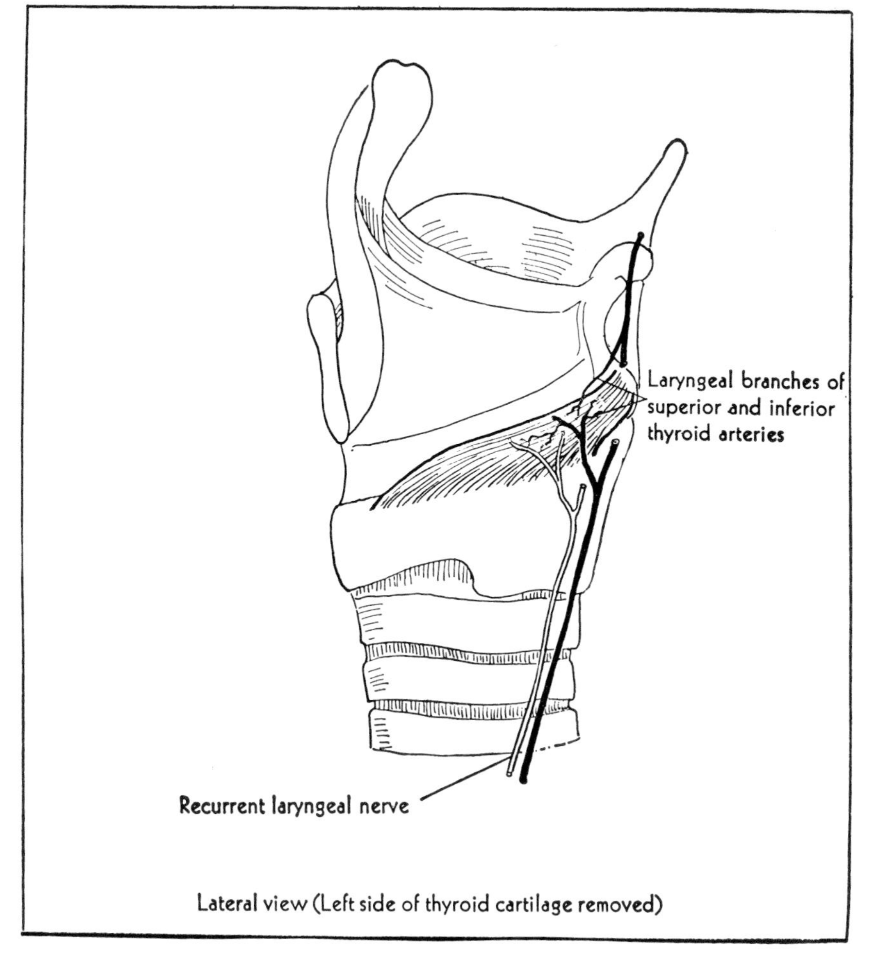

Lateral view (Left side of thyroid cartilage removed)

ORIGIN: Upper margin of arch of cricoid cartilage
INSERTION: Anterior margin of muscular process of arytenoid cartilage
FUNCTION: Closes glottis by approximating vocal folds through inward rotation of arytenoid cartilages
NERVE: Recurrent laryngeal
ARTERY: Laryngeal branches of superior and inferior thyroid

REFERENCES: GRAY	GRANT'S ATLAS
Muscle 1129	628
Nerve 1131, 941	628
Artery 581, 605	Not shown

ARYTENOIDEUS: OBLIQUUS AND TRANSVERSUS

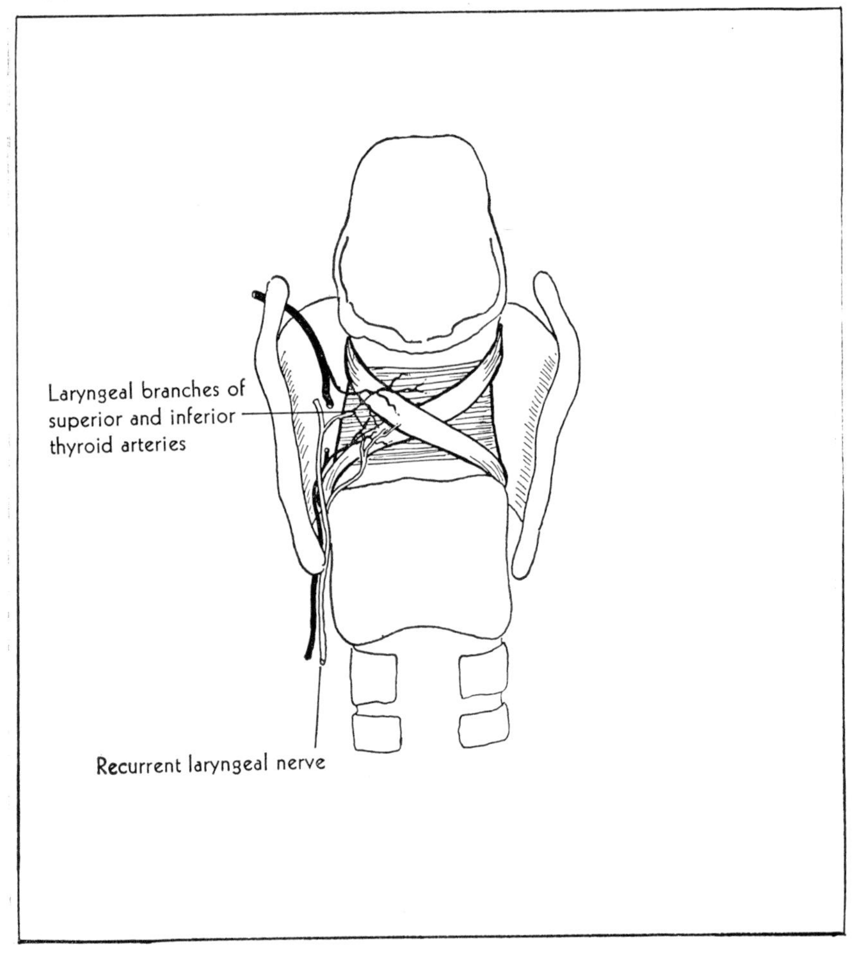

ORIGIN and INSERTION:	Transverse portion, a single muscle crossing transversely between the two arytenoid cartilages; oblique portion, more superficial, from base of one arytenoid cartilage to apex of the other
FUNCTION:	Closes glottis by approximating arytenoid cartilages
NERVE:	Recurrent laryngeal
ARTERY:	Laryngeal branches of superior and inferior thyroid

REFERENCES: GRAY	GRANT'S ATLAS
Muscle 1129	627, 632
Nerve 1131, 941	627
Artery 581, 605	Not shown

THYREOARYTENOIDEUS

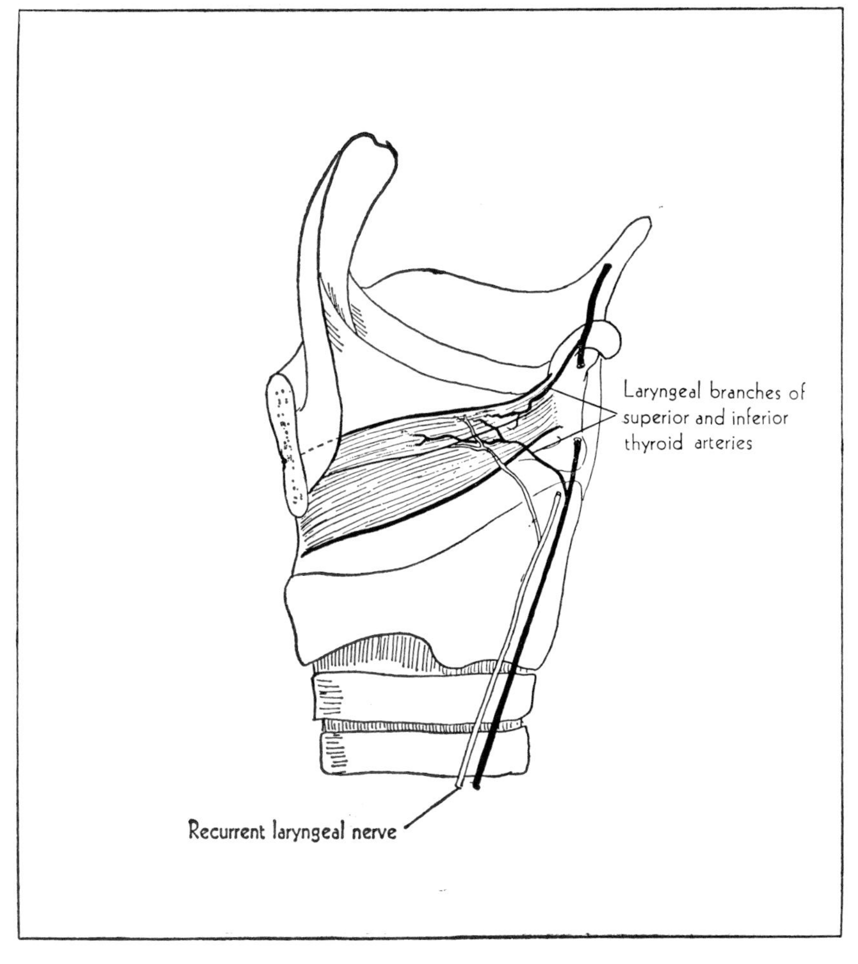

ORIGIN: Lower half of medial surface of lamina of thyroid cartilage and cricothyroid ligament

INSERTION: Base and anterior border of arytenoid cartilage

FUNCTION: Shortens and relaxes vocal folds by drawing arytenoid cartilages forward; closes glottis by inward rotation of arytenoid cartilages

NERVE: Recurrent laryngeal

ARTERY: Laryngeal branches of superior and inferior thyroid

REFERENCES:	GRAY	GRANT'S ATLAS
Muscle	1129	628
Nerve	1131, 941	628
Artery	581, 605	Not shown

VOCALIS

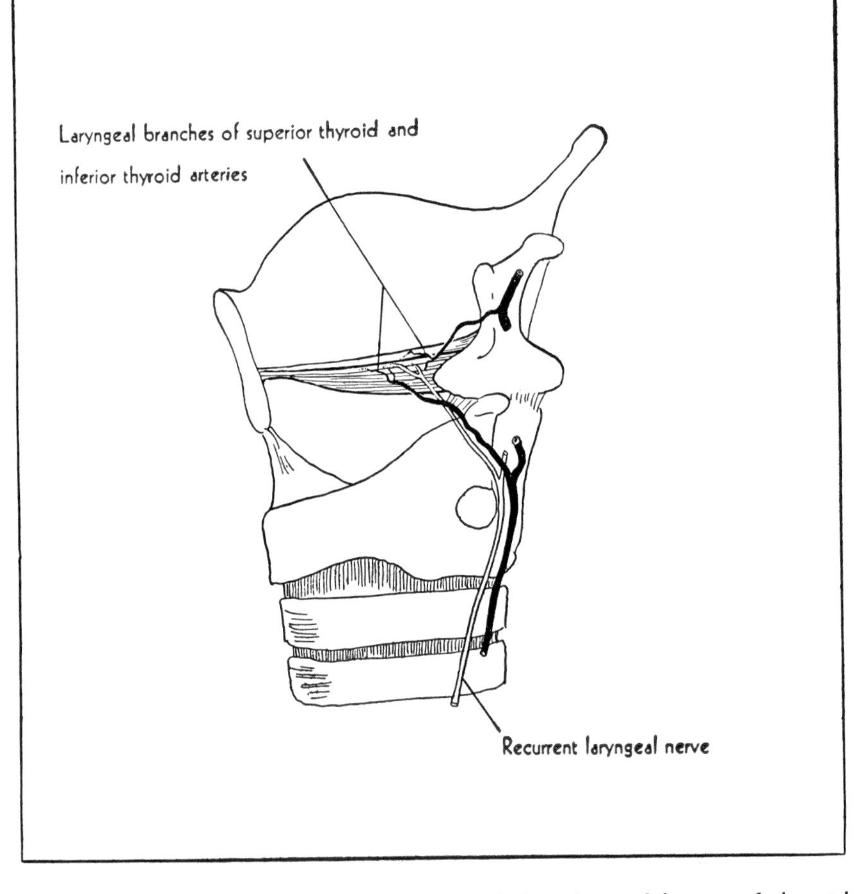

ORIGIN: With thyroarytenoid from medial surface of lamina of thyroid cartilage and cricothyroid ligament

INSERTION: Vocal process and anterior surface of arytenoid cartilage

FUNCTION: Adjusts tension of vocal ligament together with transverse arytenoid muscle; possibly controls pitch of voice

NERVE: Recurrent laryngeal

ARTERY: Laryngeal branches of superior and inferior thyroid

REFERENCES:

	GRAY	GRANT'S ATLAS
Muscle	1130	629
Nerve	1131, 941	Not shown
Artery	581, 605	Not shown

THYREOEPIGLOTTICUS

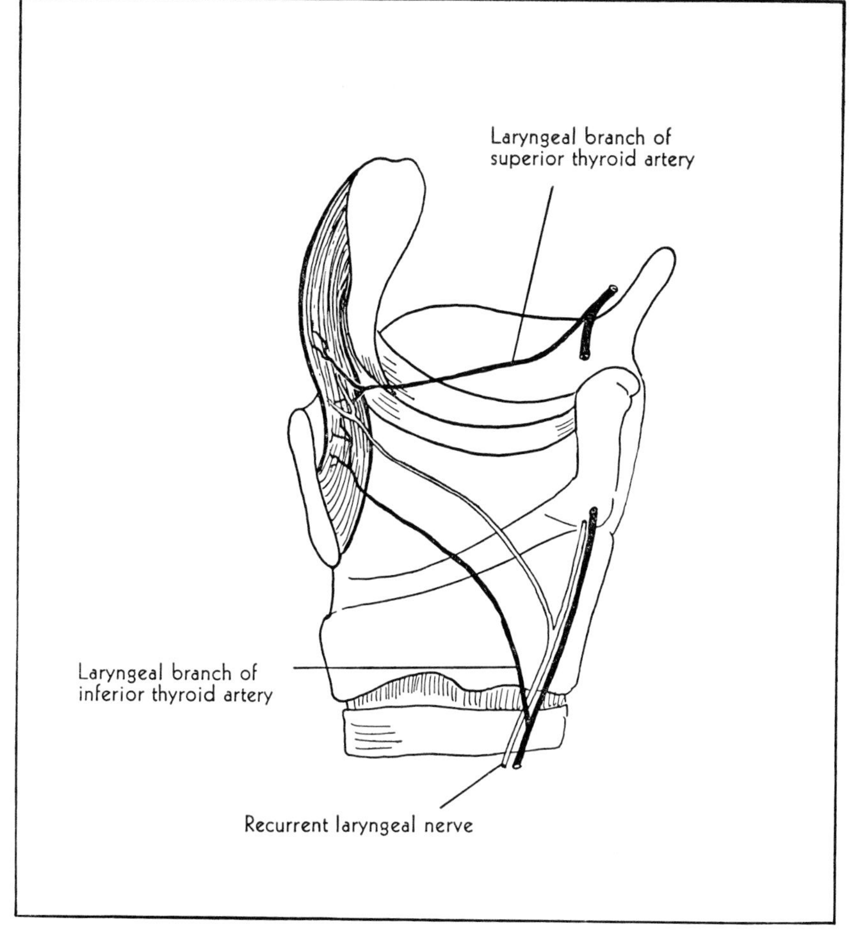

ORIGIN: Lower half of medial surface of lamina of thyroid cartilage and cricothyroid ligament
INSERTION: Lateral margin of epiglottis and aryepiglottic fold
FUNCTION: Depresses epiglottis; acts as a sphincter of the inlet of the larynx
NERVE: Recurrent laryngeal
ARTERY: Laryngeal branches of superior and inferior thyroid

REFERENCES: GRAY		GRANT'S ATLAS
Muscle	1130	628
Nerve	1131, 941	628
Artery	581, 605	Not shown

GENIOGLOSSUS

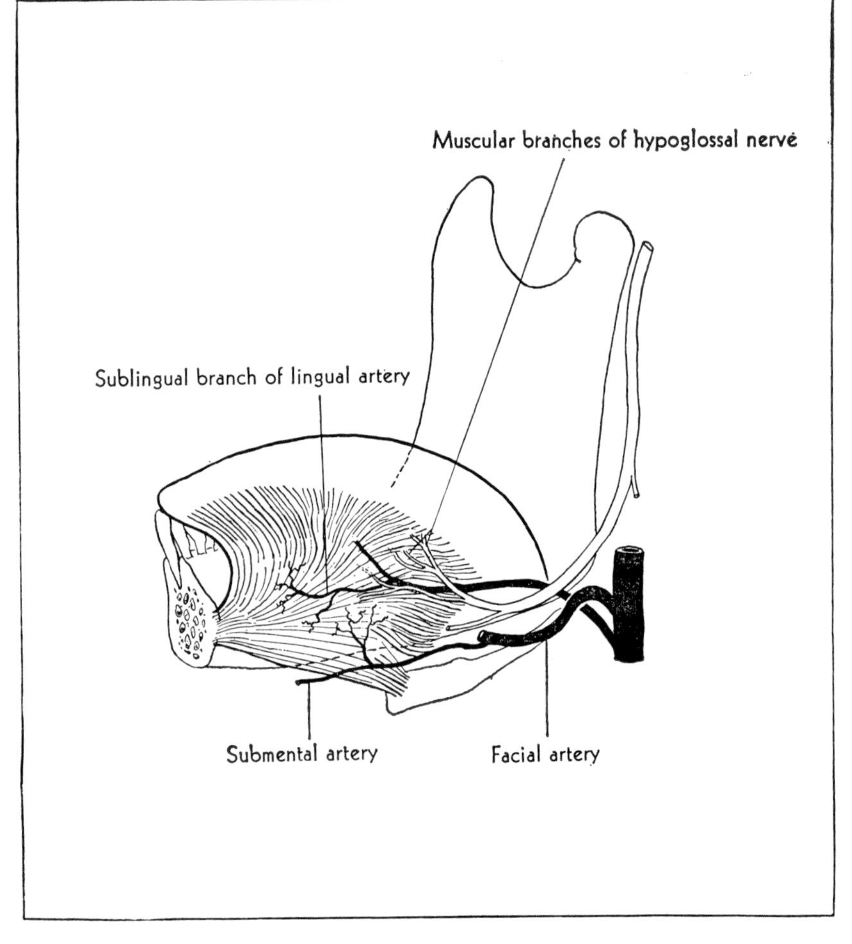

ORIGIN: This is an extrinsic tongue muscle. It arises from the upper genial tubercle of the mandible

INSERTION: Lowest fibers into body of hyoid bone, middle fibers along whole under-surface of tongue, superior fibers to tip of tongue

FUNCTION: Tongue protruded by posterior fibers, retracted by anterior fibers aided by styloglossus, depressed by genioglossus and hyoglossus

NERVE: Muscular branches of hypoglossal (see intrinsic muscles for sensory innervation of tongue) p. 107-108

ARTERY: Sublingual, submental

REFERENCES:

	GRAY	GRANT'S ATLAS
Muscle	1182	548, 592, 597
Nerve	1184, 945-946	662
Artery	582, 583	548, 549

HYOGLOSSUS *

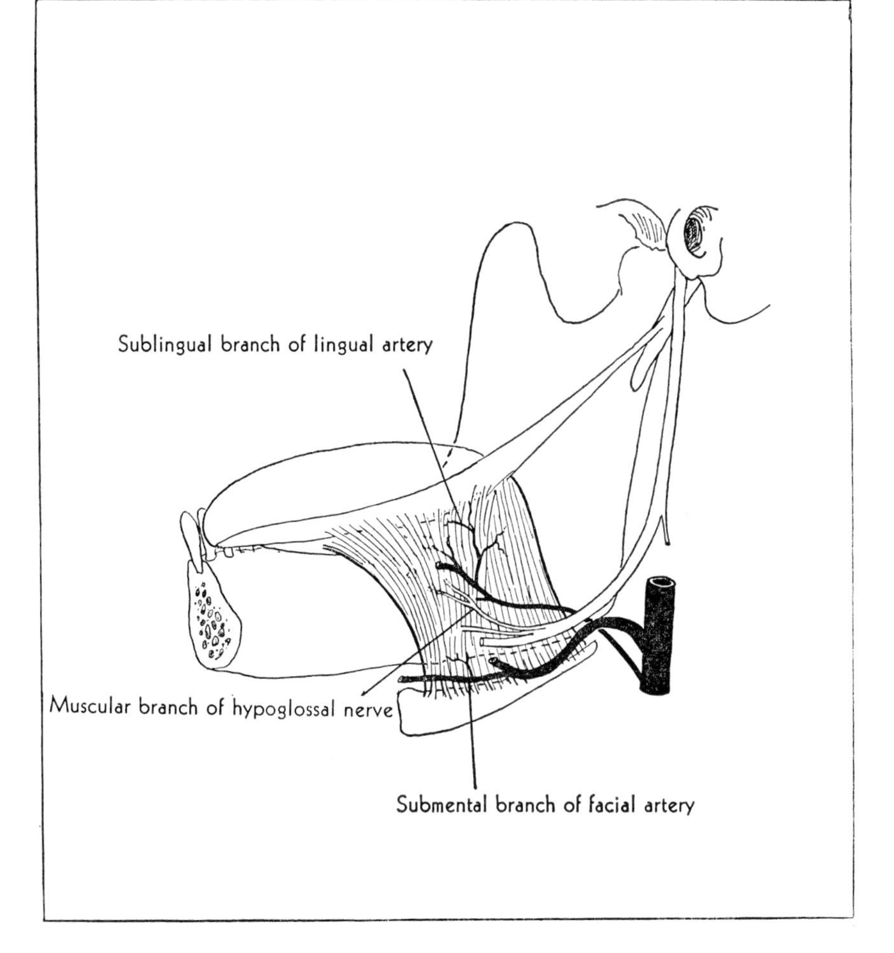

ORIGIN: This is an extrinsic tongue muscle. It arises from the sides and body of greater cornu of hyoid bone

INSERTION: Into sides of tongue, interlacing with fibers of styloglossus and longitudinalis inferior

FUNCTION: Draws down sides of tongue and with genioglossus depresses tongue

NERVE: Muscular branches of hypoglossal

ARTERY: Sublingual, submental

REFERENCES:	GRAY	GRANT'S ATLAS
Muscle	1182	546-548
Nerve	1184, 945-946	662
Artery	582, 583	548

*Chondroglossus: Gray 1182; Grant 596. This muscle arises from lesser cornu of hyoid bone; fibers pass upward to blend with intrinsic tongue muscles, between Hyoglossus and Genioglossus.

STYLOGLOSSUS

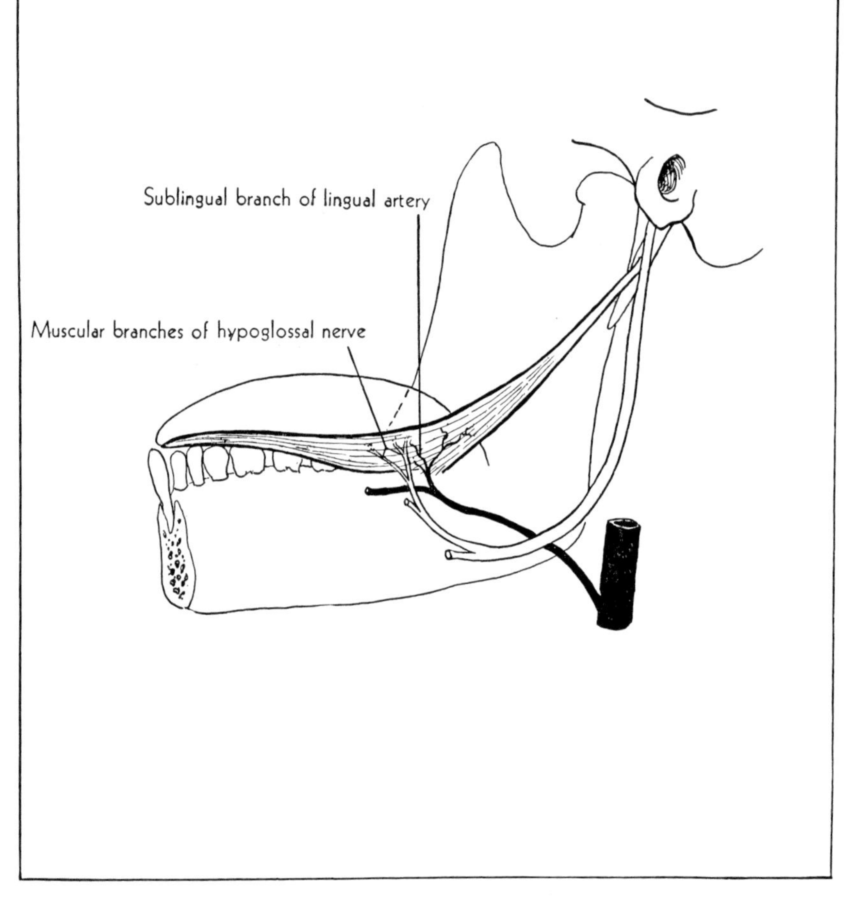

ORIGIN: This is an extrinsic muscle of the tongue. It arises from the anterior border of the styloid process

INSERTION: Into the sides of the tongue, its fibers spreading and mingling with the palatoglossus and hyoglossus

FUNCTION: Tongue retracted by styloglossus with aid of anterior fibers of genioglossus, elevated by styloglossus with aid of palatoglossus

NERVE: Muscular branches of hypoglossal

ARTERY: Sublingual

REFERENCES:

	GRAY	GRANT'S ATLAS
Muscle	1182	548, 572, 589
Nerve	1184, 945-946	662
Artery	582	Not shown

LONGITUDINALIS LINGUAE Sup. & Inf.

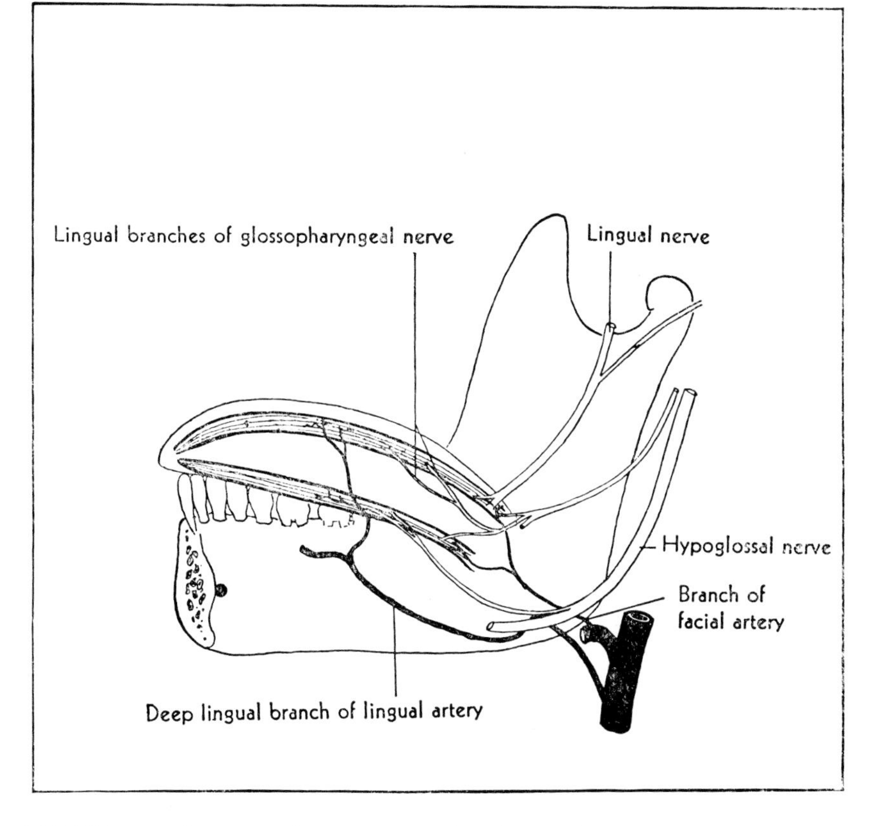

ORIGIN: Consists of superior and inferior division; intrinsic tongue muscles; 1) superior from submucous fibers beginning at the back of the tongue; 2) inferior on under surface of tongue between genio-glossus and hyo-glossus

INSERTION: 1) To tip of tongue; unites with muscle of opposite side; 2) to tip of tongue blending with styloglossus

FUNCTION: Modify shape of tongue; 1 and 2 shorten tongue, 1 turns tip and sides upward, 2 turns tip and sides downward

NERVE: Cr. N. V: Lingual branch of V^3—Sensory to anterior ⅔
Cr. N. VII: Chorda tympani branch of facial—Taste, anterior ⅔
Cr. N. IX: Glossopharyngeal—Sensory and taste, posterior ⅓
Cr. N. XII: Hypoglossal—Extrinsic and intrinsic muscles

ARTERY: Deep lingual branch of lingual artery, branches from facial

REFERENCES:	GRAY	GRANT'S ATLAS
Muscle	1184	592-595
Nerve	1184, 945-946	662
Artery	582	549.1

TRANSVERSUS AND VERTICALIS LINGUAE

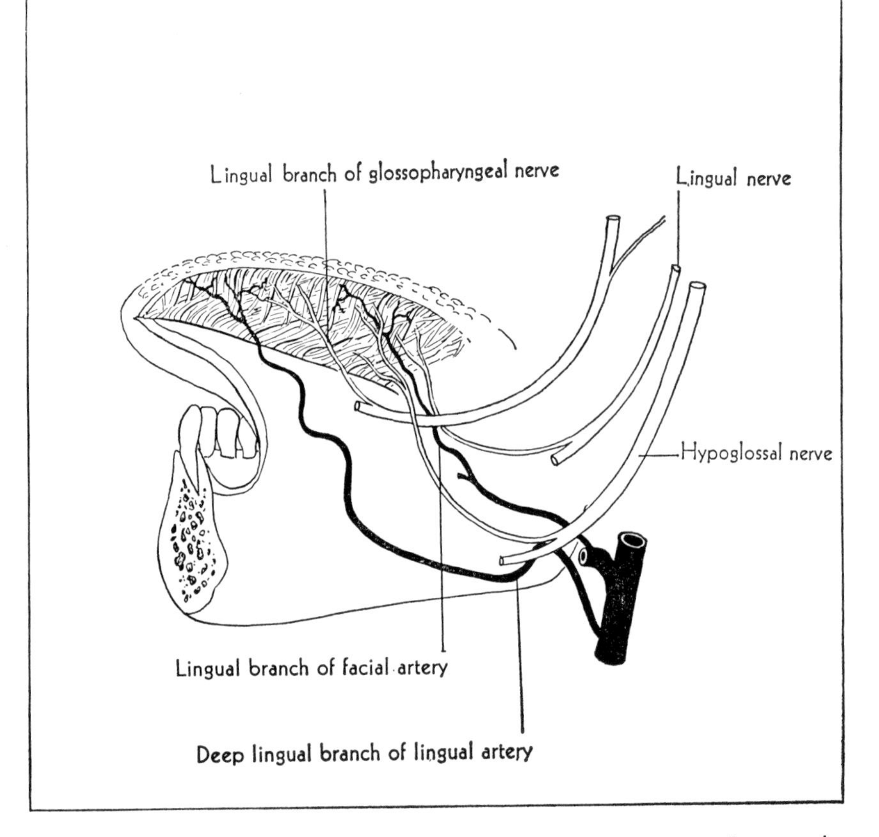

ORIGIN: These are intrinsic tongue muscles: 1) transversus from median fibrous septum, 2) verticalis from mucous membrane on dorsum of forepart of tongue

INSERTION: 1) To dorsum and sides of tongue; 2) fibers extend from dorsum to under surface of tongue

FUNCTION: Modify shape of tongue: 1) narrows and elongates tongue; 2) flattens and broadens tongue. Together they form large part of the tongue musculature

NERVE: Cr. N. V: Lingual branch of V³—Sensory to anterior ⅔
Cr. N. VII: Chorda tympani branch of facial—Taste, anterior ⅔
Cr. N. IX: Glossopharyngeal—Sensory and taste, posterior ⅓
Cr. N. XII: Hypoglossal—Extrinsic and intrinsic muscles

ARTERY: Deep lingual branch of lingual artery, branches from facial

REFERENCES:	GRAY	GRANT'S ATLAS
Muscle	1184	592-595
Nerve	1184, 945-946	662
Artery	582	549.1

LEVATOR VELI PALATINI (Levator palati)

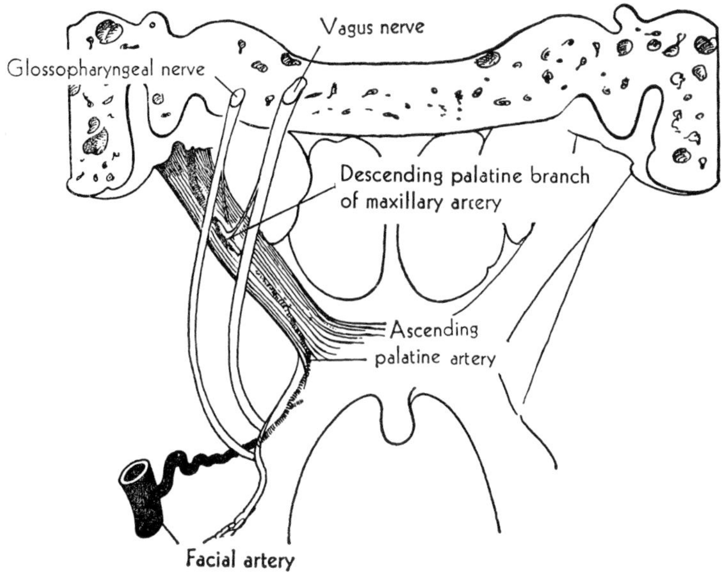

ORIGIN: Lower surface of petrous portion of temporal bone; medial side of cartilage of auditory tube

INSERTION: Fibers extend downward and medially to mid-line and join those of opposite side

FUNCTION: Raises soft palate in swallowing

NERVE: Pharyngeal plexus, formed by contributions from sympathetic glossopharyngeal, vagus and cranial portion of accessory

ARTERY: Ascending palatine branch of facial; descending palatine branch of maxillary

REFERENCES:	GRAY	GRANT'S ATLAS
Muscle	1194	575, 588, 639
Nerve	1194, 935, 938, 944	Not shown
Artery	583, 591	576, 577

TENSOR VELI PALATINI (Tensor palati)

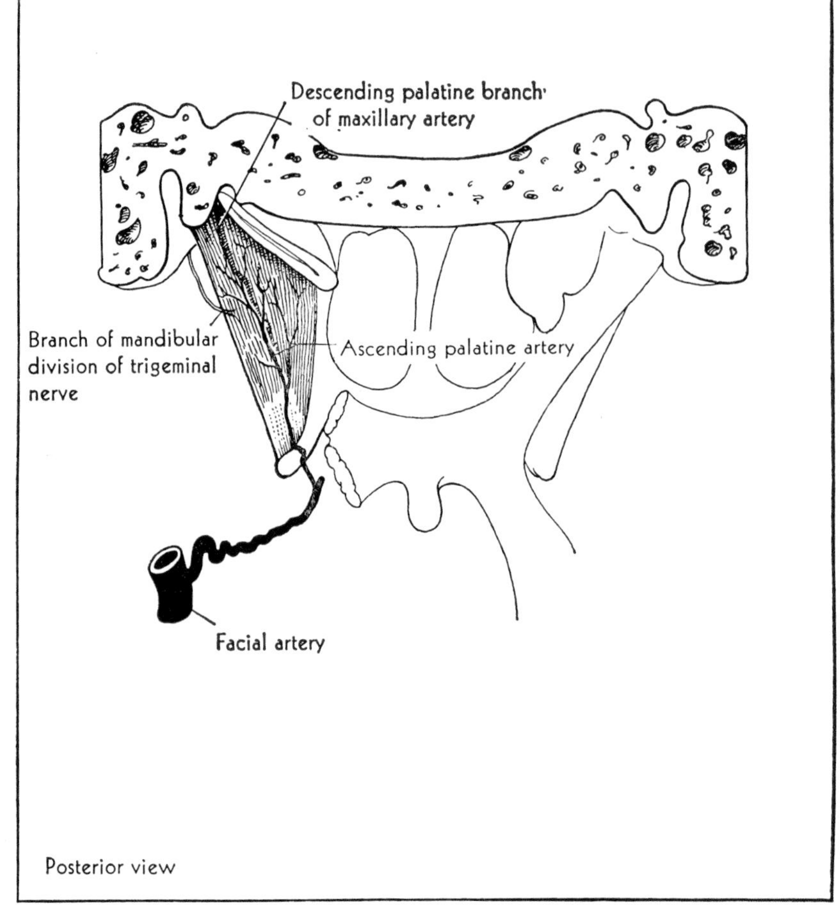

Posterior view

ORIGIN: Scaphoid fossa and spine of sphenoid bone; lateral sides of membranous and cartilaginous portions of auditory tube

INSERTION: Tendon winds around pterygoid hamulus and into aponeurosis of soft palate; posterior part of palatine bone

FUNCTION: Tenses soft palate; opens auditory tube during swallowing

NERVE: Small branch from mandibular division of trigeminal

ARTERY: Ascending palatine branch of facial, descending palatine branch of maxillary

REFERENCES:

	GRAY	GRANT'S ATLAS
Muscle	1194	575, 577, 588, 590
Nerve	1194, 920, 934	654
Artery	583, 591	582

MUSCULUS UVULAE

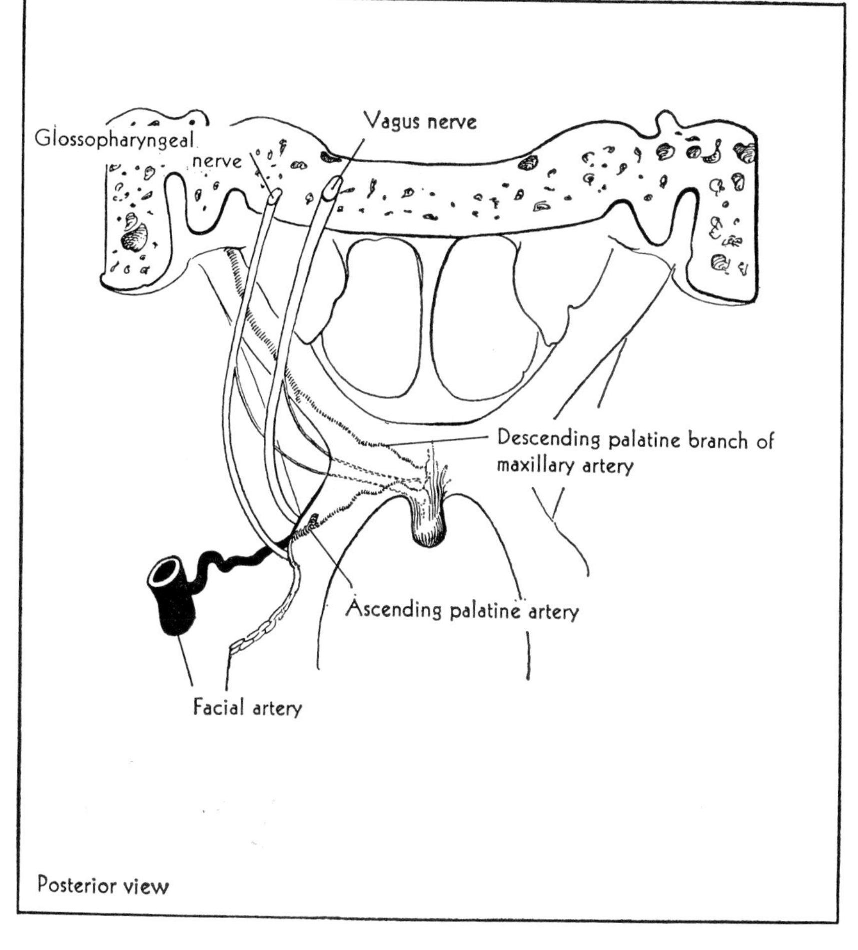

ORIGIN: Posterior nasal spine and palatine aponeurosis
INSERTION: Mucous membrane of uvula
FUNCTION: Raises uvula
NERVE: Pharyngeal plexus, formed by contributions from sympathetic, glossopharyngeal, vagus and cranial portion of accessory
ARTERY: Ascending palatine branch of facial; descending palatine branch of maxillary

REFERENCES:	GRAY	GRANT'S ATLAS
Muscle	1194	588
Nerve	1194, 935, 938, 944	Not shown
Artery	583, 591	Not shown

PALATOGLOSSUS (Glossopalatinus)

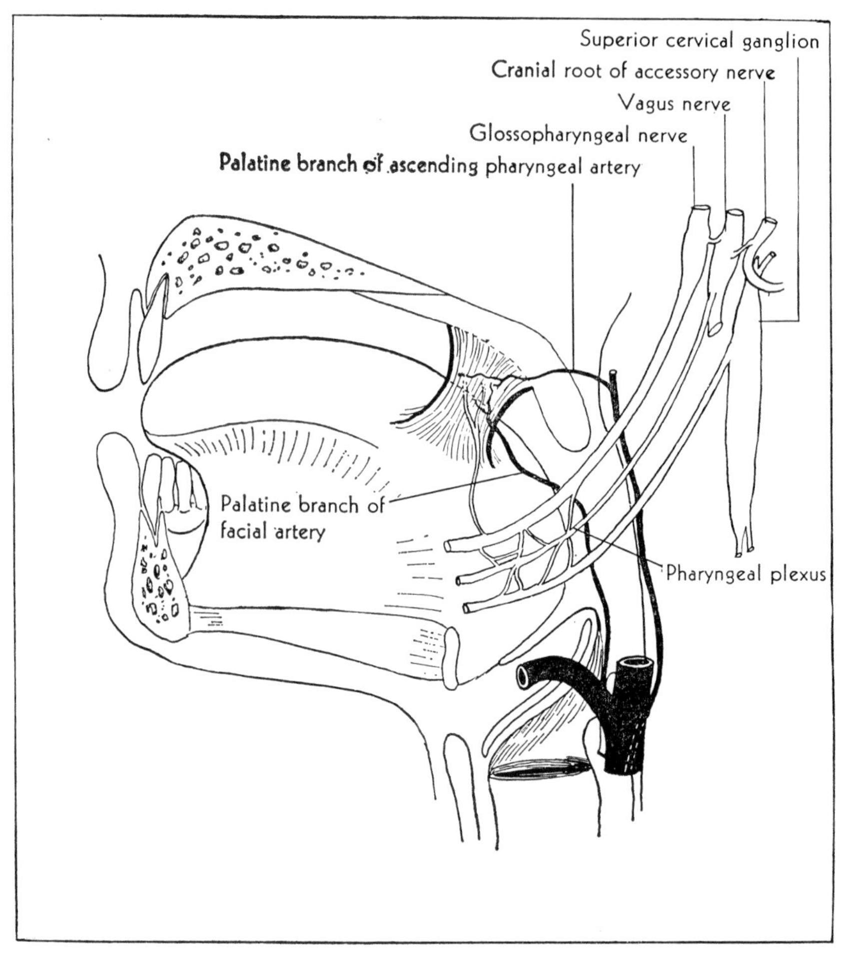

ORIGIN: Anterior surface of soft palate

INSERTION: Dorsum and side of tongue blending with styloglossus and transversus linguae

FUNCTION: Narrows fauces and elevates back of tongue

NERVE: Pharyngeal plexus, formed by contributions from sympathetic, glossopharyngeal, vagus and cranial portion of accessory

ARTERY: Ascending palatine branch of facial, palatine branch of ascending pharyngeal

REFERENCES:	GRAY	GRANT'S ATLAS
Muscle	1194	588, 589
Nerve	1194, 935, 938, 944	593
Artery	581, 583	Not shown

PALATOPHARYNGEUS (Pharyngopalatinus)

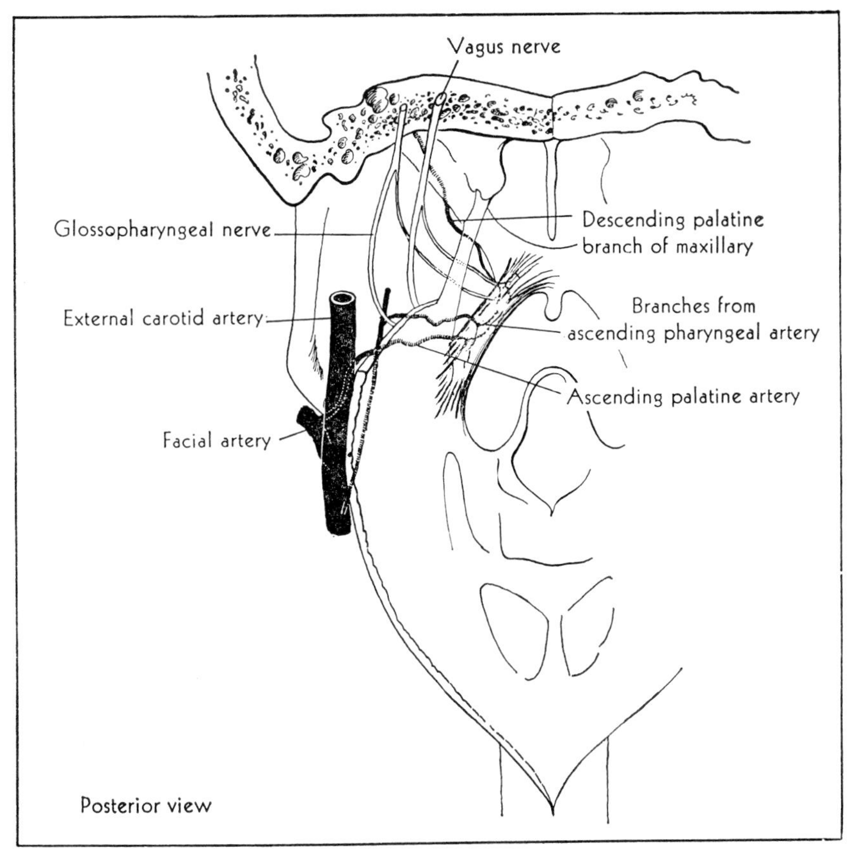

ORIGIN: Extends from soft palate to pharyngeal wall, with mucous membrane forms pharyngopalatine arch; fibers arise from soft palate in 2 layers, the posterior-superior in contact with mucous membrane joins opposite muscle in mid-line; antero-inferior layer passes laterally and downward, joins opposite muscle in mid-line

INSERTION: Posterior border of thyroid cartilage and aponeurosis of pharynx

FUNCTION: Narrows oro-pharyngeal isthmus, elevates pharynx, shuts off nasopharynx

NERVE: Pharyngeal plexus, formed by contributions from sympathetic, glossopharyngeal, vagus and cranial portions of accessory

ARTERY: Twigs from ascending palatine branch of facial, descending palatine branch of maxillary, palatine branch of ascending pharyngeal

REFERENCES: GRAY		GRANT'S ATLAS
Muscle	1194	588, 589
Nerve	1194, 935, 938, 944	Not shown
Artery	581, 583, 591	Not shown

CONSTRICTOR PHARYNGIS INFERIOR

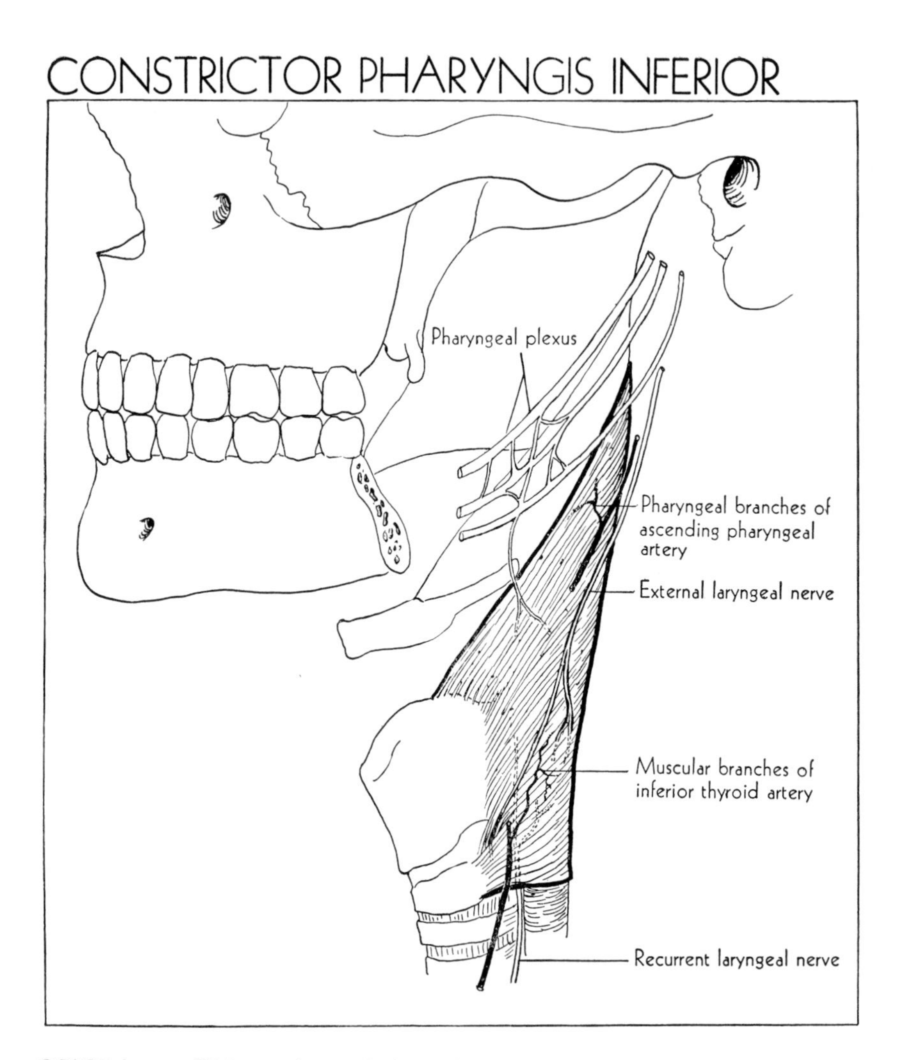

ORIGIN: Oblique line of thyroid cartilage, side of cricoid cartilage, posterior border of crico-thyroid muscle

INSERTION: Posterior median raphé of pharynx, upper fibers ascend and overlap middle constrictor, lowest fibers are horizontal and blend with circular muscular fibers of esophagus

FUNCTION: Contracts pharynx as in swallowing

NERVE: Pharyngeal plexus and external and recurrent laryngeal branches of vagus and cranial root of accessory

ARTERY: Small muscular branches of inferior thyroid; pharyngeal branches of ascending pharyngeal artery

REFERENCES:	GRAY	GRANT'S ATLAS
Muscle	1197	572, 626
Nerve	1199, 935, 938, 941, 944	626
Artery	581, 605	Not shown

CONSTRICTOR PHARYNGIS MEDIUS

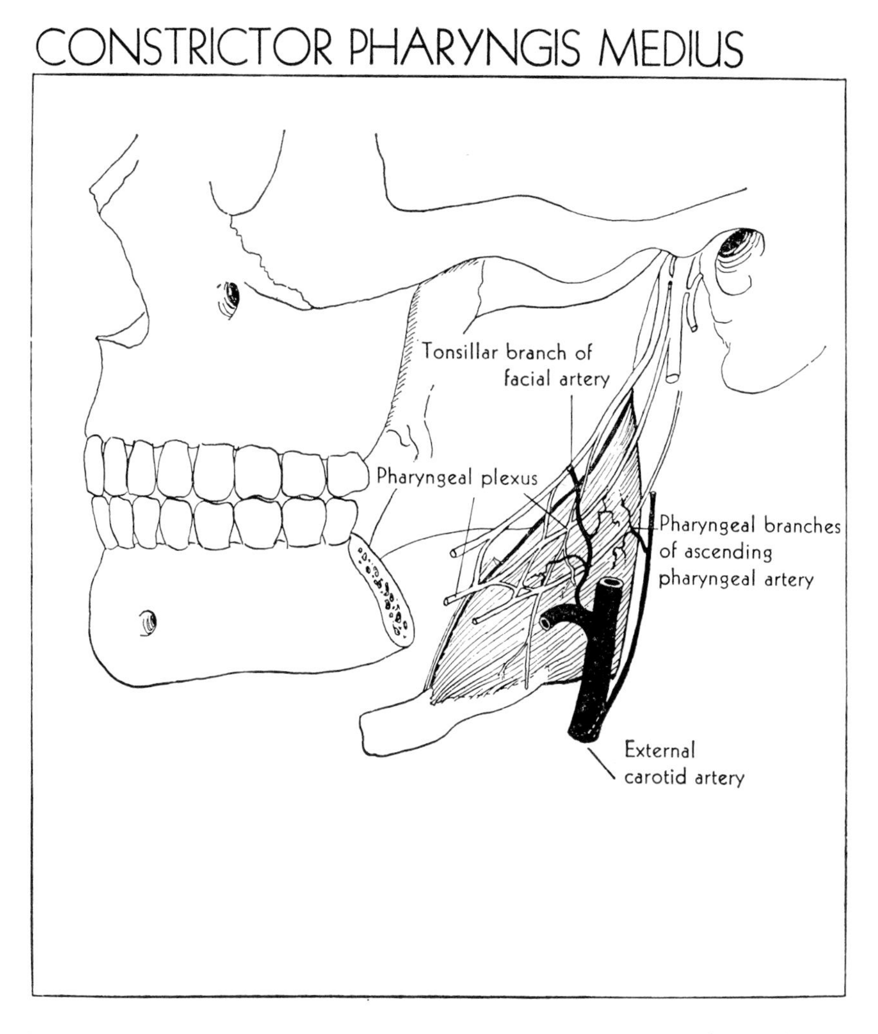

ORIGIN: Both horns of hyoid bone, stylohyoid ligament, fibers diverge from origin; upper ascend and overlap constrictor superioris, middle ones run transversely, lower pass beneath constrictor inferior

INSERTION: Into posterior median raphé, blending in mid-line with muscle of opposite side

FUNCTION: Contracts pharynx as in swallowing

NERVE: Pharyngeal plexus

ARTERY: Small pharyngeal branches of ascending pharyngeal; tonsillar branches of facial

REFERENCES:	GRAY	GRANT'S ATLAS
Muscle	1197	572, 589, 590
Nerve	1199, 935, 938, 944	660
Artery	581, 583	589

CONSTRICTOR PHARYNGIS SUPERIOR

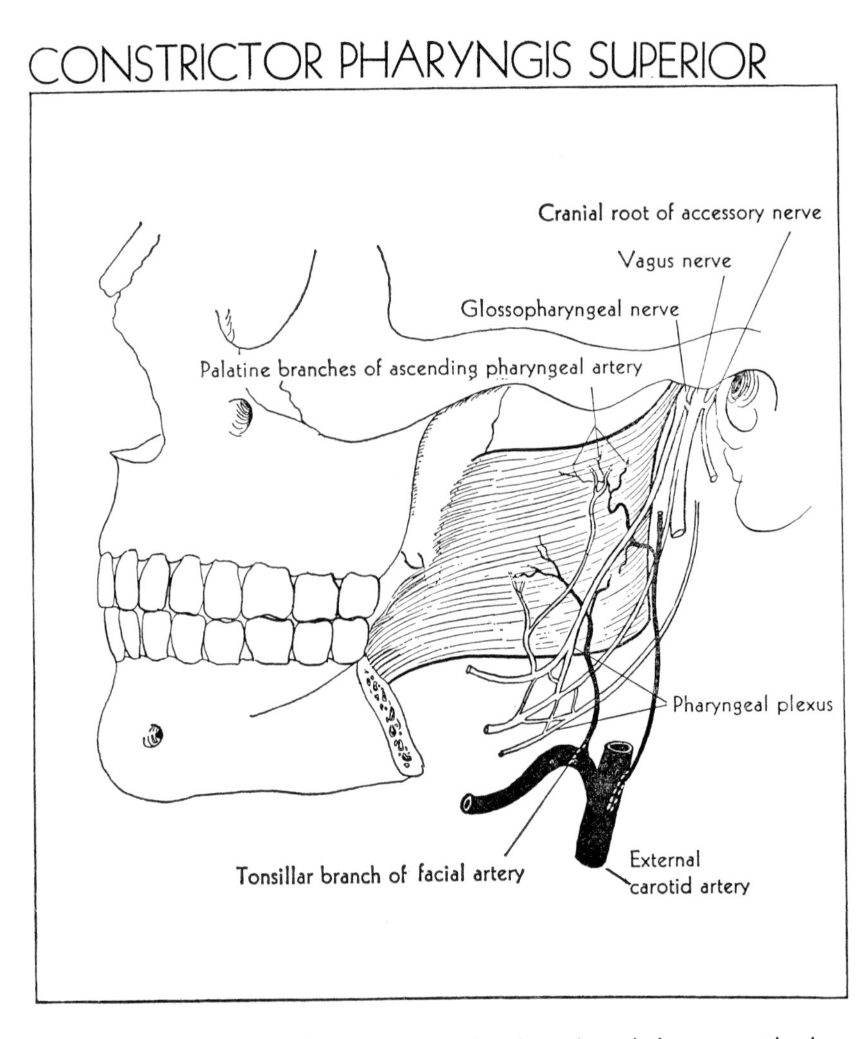

ORIGIN: Lower third of posterior border of medial pterygoid plate, pterygo-mandibular ligament, alveolar process of mandible above posterior end of mylohyoid line, side of tongue

INSERTION: Median raphé on posterior wall of pharynx, by aponeurosis to pharyngeal tubercle on occipital bone

FUNCTION: Contracts pharynx in swallowing

NERVE: Cranial root of accessory; vagus, through pharyngeal plexus

ARTERY: Tonsillar branch, facial artery; palatine branch of ascending pharyngeal

REFERENCES: GRAY	GRANT'S ATLAS
Muscle 1197	572, 589, 590
Nerve 1199, 935, 938, 944	660
Artery 581, 583	590

STYLOPHARYNGEUS

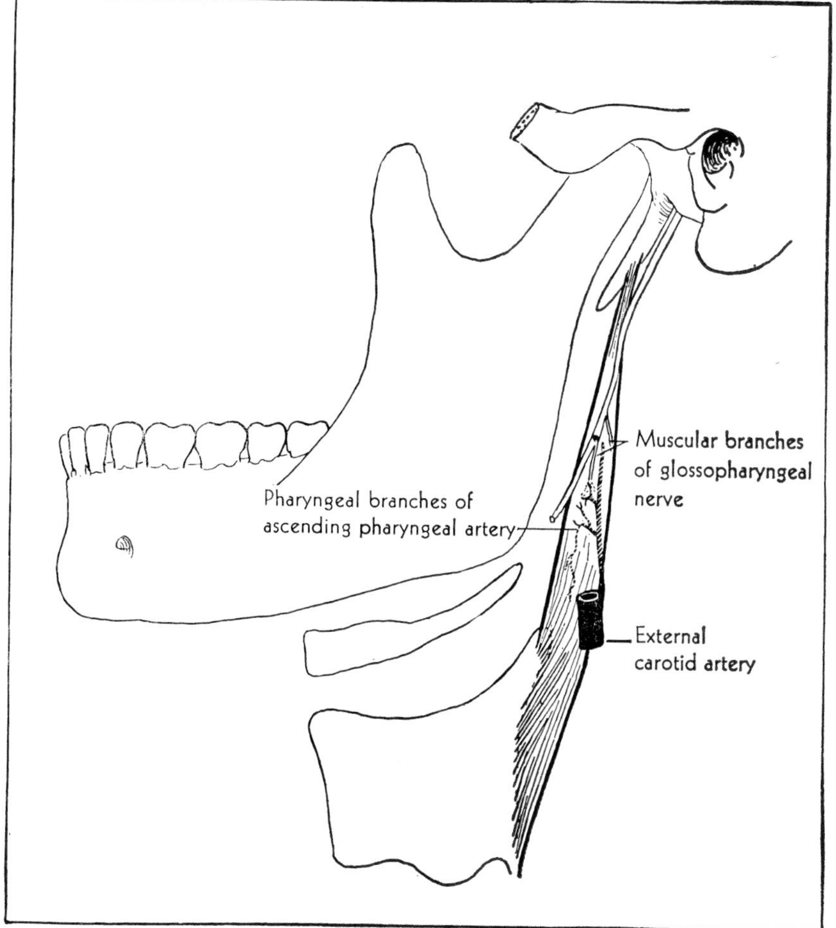

ORIGIN: Medial side of root of styloid process, passes downward between external and internal carotid arteries
INSERTION: Superior and posterior borders of thyroid cartilage; some fibers mingle with those of constrictors of pharynx
FUNCTION: Raises and dilates pharynx
NERVE: Muscular branch of glossopharyngeal
ARTERY: Pharyngeal branches of ascending pharyngeal

REFERENCES: GRAY	GRANT'S ATLAS
Muscle 1198	549, 572, 577, 589
Nerve 1199, 936	659
Artery 581	Not shown

SALPINGOPHARYNGEUS

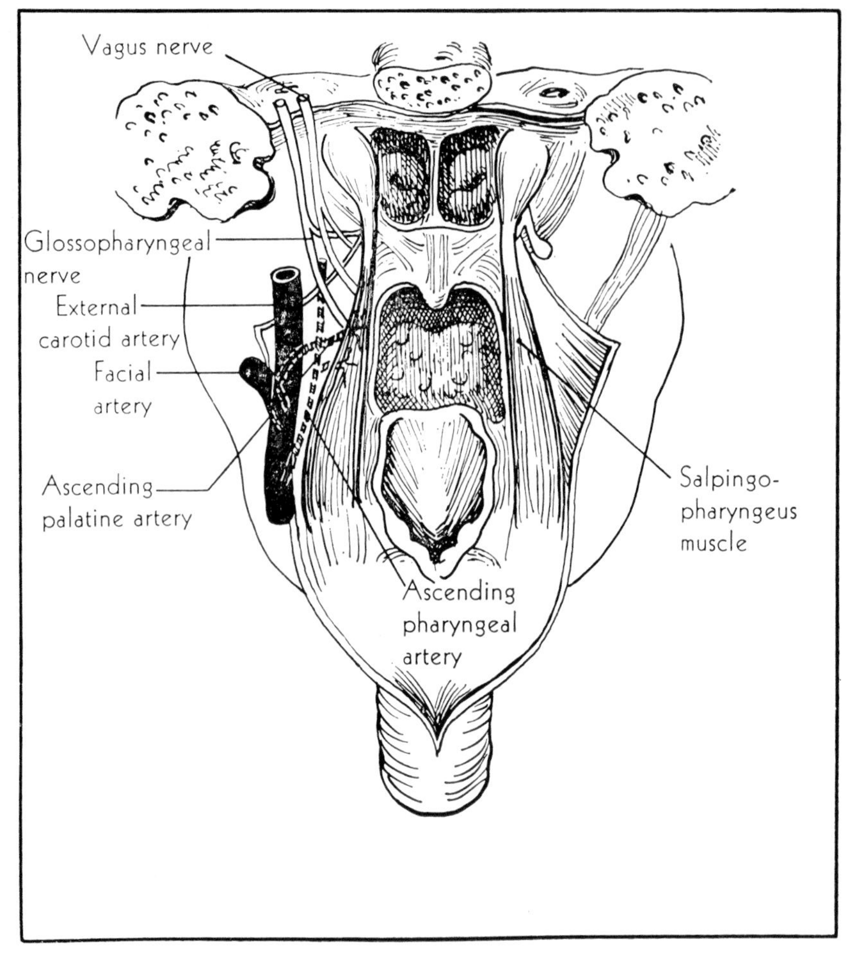

ORIGIN: Pharyngeal end of auditory tube near its orifice
INSERTION: Blends with posterior fasciculus of palatopharyngeus
FUNCTION: Raises nasopharynx
NERVE: Pharyngeal plexus
ARTERY: Twigs of ascending palatine branch of facial, descending palatine branch of maxillary, palatine branch of ascending pharyngeal

REFERENCES:	GRAY	GRANT'S ATLAS
Muscle	1198	588
Nerve	1199, 935, 938	Not shown
Artery	581, 583, 591	Not shown

MOTOR POINTS

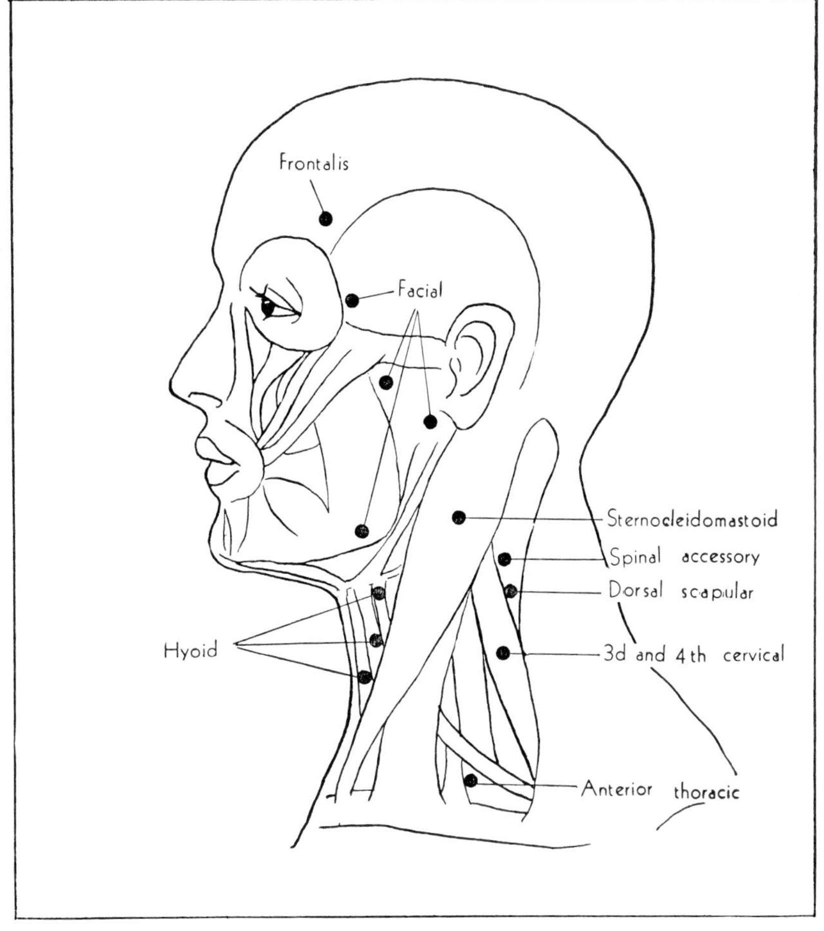

Temporal, zygomatic and buccal branches of the facial nerve supply the muscles of facial expression. The approximate positions of these main facial nerve branches are indicated in the diagram. The infra-hyoidean muscles are supplied by the hypoglossal nerve and the ansa hypoglossi. The diagram is based on the Fischer Electro-Diagnostic Chart of this region.

FORAMINA OF THE SKULL (NORMA BASALIS)

Foramen	*Structures transmitted*
Sulcus tubae auditivae	Lodges cartilaginous part of auditory tube
Hypoglossal canal (arrow)	Hypoglossal nerve; a meningeal artery
F. magnum	Spinal cord (medulla oblongata); spinal accessory nerves; vertebral arteries; anterior and posterior spinal arteries; occipitoaxial ligament
F. incisivum	Descending septal artery; Nasopalatine nerves
Greater palatine foramen	Greater palatine vessels and nerve
Inferior orbital fissure	Maxillary division of trigeminal nerve; zygomatic branch of trigeminal nerve; filaments from pterygopalatine branch of the maxillary nerve; infraorbital vessels; vein between inferior ophthalmic vein and the pterygoid venous plexus
Lesser palatine foramen	Lesser palatine nerves
F. lacerum	Closed inferiorly by a fibrocartilaginous plate which contains the auditory tube; upper part traversed by the internal carotid artery.
F. ovale	Mandibular division of trigeminal nerve
F. spinosum	Middle meningeal vessels
Carotid canal	Internal carotid artery
Stylomastoid foramen	Facial nerve; Stylomastoid artery
Jugular foramen	Beginning of internal jugular vein; cranial nerves IX, X, and XI
Condyloid foramen	Vein from transverse sinus
Mastoid foramen	Vein to transverse sinus; Branch of occipital artery to dura mater

FORAMINA OF THE SKULL (NORMA BASALIS)

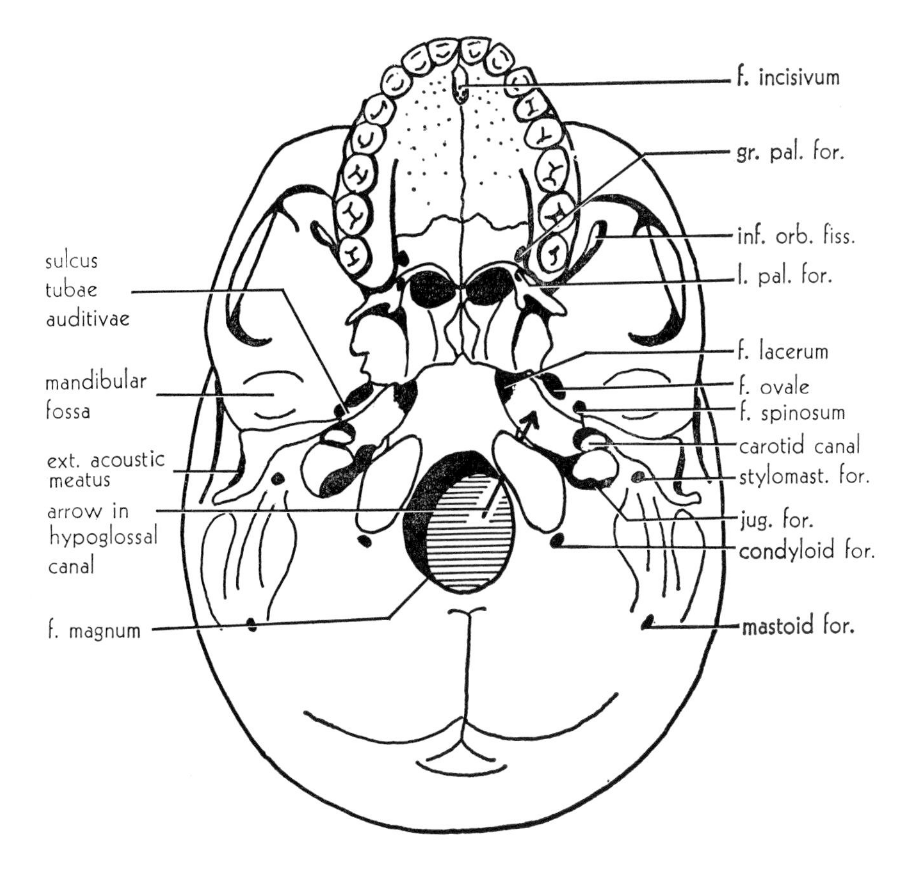

FORAMINA OF THE SKULL (BASE, INTERIOR)

Foramen	*Structures transmitted*
Nasal slits	Anterior ethmoidal nerves and vessels
Superior orbital fissure	Cranial nerves III, IV, ophthalmic division of V, VI; superior ophthalmic vein; branch from lacrimal artery to dura mater
F. lacerum	Upper part traversed by internal carotid artery
Internal acoustic meatus	Facial nerve Stato-acoustic nerve
Hypoglossal canal	Hypoglossal nerve; a meningeal artery
F. magnum	Spinal cord (medulla oblongata); spinal accessory nerves; vertebral arteries; anterior and posterior spinal arteries; occipitoaxial ligament
F. cecum	Occasional small vein
Cribriform plate	Olfactory nerves
Optic canal	Optic nerve; ophthalmic artery
F. rotundum	Maxillary division of trigeminal nerve
F. ovale	Mandibular division of trigeminal nerve
F. spinosum	Middle meningeal vessels
Jugular foramen	Beginning of the internal jugular vein; cranial nerves IX, X, XI
Condyloid foramen	Vein from transverse sinus
Mastoid foramen	Vein to transverse sinus; branch of occipital artery to dura mater

FORAMINA OF THE SKULL (BASE, INTERIOR)

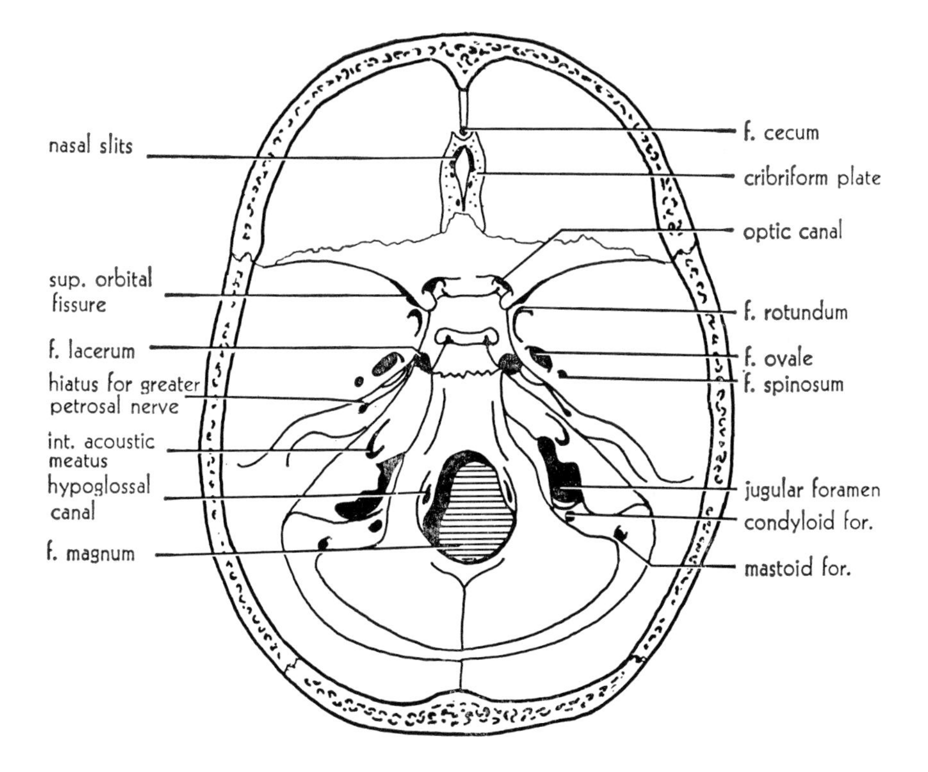

INDEX

ARTERIES

Alveolar, inferior, mylohyoid branch, 38
Auricular, posterior, 14
Occipital branch, 12, 35
Stylomastoid branch, 96
Carotid, common, 43, 44, 45
External, 15, 17, 18, 19, 29, 30, 31, 35, 38, 39, 40, 116, 117, 118
Internal, 92
Cervical, ascending, 44, 45, 48, 49, 50
Transverse,
Superficial branch, 50, 51
Costocervical, 59, 65, 66
Epigastric, inferior, 73, 74, 76
Cremasteric branch, 75
Muscular branch, 77, 78
Superior, 65, 73, 74
Facial, 15, 17, 20, 23, 24, 25, 26, 27, 28, 36, 104, 113, 118
Alar branch, 20
Angular branch, 15, 17, 21
Ascending palatine branch, 109, 110, 111, 112, 118
Labial branch, 18, 20, 21, 22, 23, 24, 25, 26, 27, 28
Lateral nasal branch, 17, 18, 19
Lingual branch, 108
Muscular branch, 29, 31, 32, 37, 107
Septal branch, 18, 20
Submental branch, 34, 38, 105
Tonsillar, branch, 115, 116
Transverse, 31
Gluteal, inferior muscular branch, 80, 81
Hemorrhoidal (See Rectal)
Hypogastric, 81
Iliac circumflex, deep, 74, 76
Illolumbar, 79
Infraorbital, 91, 94
Intercostal, 53, 54, 55, 65, 66
Muscular branch, 57, 58, 60, 67
Posterior, 56, 59, 69, 70, 71
Labial, inferior, 25, 26, 27, 28
Superior, 21, 22, 23, 24, 28
Lingual, hyoid branch, 40, 43
Deep lingual, 107, 108
Sublingual branch, 38, 39, 104, 105, 106
Lumbar, muscular branch, 57, 59, 60
Mammary, internal (See Thoracic, internal)
Masseteric, 31
Maxillary, buccal branch, 29
Descending palative branch, 109, 110, 111
Pterygoid branch, 32, 33
Maxillary, external, (See Facial)
Internal, (See Maxillary)
Meningeal, middle, 96,
Superior tympanic branch, 95
Musculophrenic, 65, 66
Nasal, lateral, 17, 18, 20
Occipital, 12
Descending branch, 61–64
Muscular branch, 36, 37, 47, 51
Sternocleidomastoid branch, 35
Ophthalmic, inferior muscular branch, 91, 92, 94
Superior muscular branch, 90, 91, 93
Pharyngeal, ascending, 44–47, 118
Palatine branch, 112, 116
Pharyngeal branch, 114, 115, 117
Phrenic, inferior, 72
Posterior auricular, 14
Occipital branch, 12
Pudendal, internal, 80, 82, 83, 86, 87, 89
Muscular branch, 81

ARTERIES—Continued
Perineal branch, 84, 85
Rectal, inferior, 80, 89
Transverse perineal, 89
Subclavian, 48, 49, 50
Submental, 36, 104
Superficial temporal, 13
Ant. br., 16
Frontal branch, 12
Parietal branch, 14
Zygomatic branch, 15, 16
Supraorbital, 90
Suprascapular, sterno-cleidomastoid branch, 35
Suprasternal branch, 34
Temporal, anterior, 30
Deep, 30
Middle, 30
Posterior, 30
Superficial, 30
Anterior branch, 16
Frontal branch, 12
Parietal branch, 14
Zygomatic branch, 15, 16
Thoracic, internal, 65
Infercostal branch, 66
Sternal branch, 68
Thyroid, inferior, 49, 50
Laryngeal branch, 97–103
Muscular branch, 114
Superior, circothyroid branch, 41, 97
Laryngeal branch, 98–103
Sternocleidomastoid, branch, 35, 40, 43
Tympanic, anterior, 96
Vertebral, 44
Muscular branch, 45, 46, 47, 61, 62, 63, 64

CHARTS
Foramina of skull (norma basalis), 120–121
Foramina of skull (base, interior), 122–123

MUSCLES
Arytaenoideus, 100
Auriculares: anterior, superior, posterior, 14
Buccinator, 29
Bulbocavernosus, 83, 84
Caninus, 22 (Levator anguli oris)
Chondroglossus, 105
Coccygeus, 81
Compressor naris, 18 (Nasalis)
Constrictor pharyngis inferior, 114
Medius, 115
Superior, 116
Corrugator, 16
Cutis ani, 89
Cremaster, 75
Cricoarytaenoideus lateralis, 99
Posterior, 98
Cricothyreoideus, 97
Depressor anguli oris, 26 (Triangularis)
Labii inferioris, 25 (Quadratus)
Septi, 20
Diaphragm, 72
Digastricus, 36
Dilatator naris, 19 (Nasalis)
Epicranius, 12
Erector spinae, 52
Genioglossus, 104
Geniohyoideus, 39
Glossopalatinus, 112 (Palatoglossus)
Hyoglossus, 105
Iliocostalis: cervicis, thoracis, lumborum, 53
Intercostales, externi, 65
Interni, 66
Interspinales, 59
Intertransversarii, 60
Ischiocavernosus, 85
Ischio-coccygeus, 81
Levator anguli oris, 22 (caninus)
Ani, 80
Labii superioris alaequae nasi, 21 (Quadratus)

MUSCLES—Continued
Palati, 109 (L. veli palatini)
Palpebrae superioris, 90
Veli palatini, 109
Levatores costarum, 69
Longissimus capitis, cervicis, thoracis, 54
Longitudinalis linguae, 107
Longus capitis, 45
Cervicis, 44 (L. colli)
Longus colli, 44 (L. cervicis)
Masseter, 31
Mentalis, 27
Multifidus, 57
Musculus uvulae, 111
Mylohyoideus, 38
Nasalis, 18 (Compressor naris) (Dil. naris) 20
Obliquus capitis inferior, 63
Superior, 64
Externus abdominus, 73
Internus abdominis, 74
Oculi inferior, 94
Superior, 93
Occipito-frontalis, 12 (Epicranius)
Omohyoideus, 43
Orbicularis oculi, 15
Oris, 28
Palatoglossus, 112 (Glossopalatinus)
Palatopharyngeus, 113
Pharyngopalatinus, 113
Platysma, 34
Procerus, 17 (Pyramidalis nasi)
Pterygoideus, lateralis, 33
Medialis, 32
Pyramidalis, 78
Nasi, 17 (Procerus)
Quadratus labii inferioris, 25 (Depressor)
Superioris, 21 (Levator)
Quadratus lumborum, 79
Recti: lateralis and medialis, 92
Superior and inferior, 91
Rectus abdominis, 77
Capitis anterior, 46
Lateralis, 47
Posterior major, 61
Minor, 62
Risorius, 24
Rotatores, 58
Sacrospinalis, 52 (Erector spinae)
Salpingopharyngeus, 118
Scalenus anterior, 48
Medius, 49
Posterior, 50
Semispinalis: capitis, cervicis, Thoracis, 56
Serratus posterior inferior, 71
Superior, 70
Spinalis: capitis, cervicis, thoracis, 55
Sphincter ani externus and internus, 89
Urethrae, 88
Splenius capitis and cervicis, 51
Stapedius, 96
Sternocleidomastoideus, 35
Sternohyoideus, 40
Sternothyreoideus, 41
Styloglossus, 106
Stylohyoideus, 37
Stylopharyngeus, 117
Subcostales, 67
Temporalis, 30
Temporoparietalis, 13
Tensor palati, 110
Tympani, 95
Veli palatini, 110
Thyreoarytaenoideus, 101
Thyreoepiglotticus, 103
Thyreohyoideus, 42
Transversus abdominis, 76
Linguae, 108
Perinei profundus, 86, 87
Perinei superficialis, 82
Thoracis, 68
Triangularis, 26 (Depressor anguli oris)
Verticalis linguae, 108
Vocalis, 102
Zygomaticus major, 23

NERVES

Abducens, 92
Accessory, cranial root, 116
  Spinal root, muscular branch, 35
Alveolar, inferior, mylohyoid branch, 36, 38
Ansa cervicalis, 40, 41, 43
Cervical, 56
  1st, 46, 47, 61, 62, 63, 64
  2nd, 35, 40, 41, 43, 46, 47, 63
  3d, 40, 41, 43, 49
  4th, 49
  Middle and lower, 51
  Muscular branch, 35, 44, 45, 48, 50
Dorsal, of clitoris, 87
  of penis, 86
Facial, 96
  Buccal branch, 18, 19, 20, 21, 22, 23, 24, 26, 28, 29
  Cervical branch, 34
  Digastric branch, 36
  Mandibular branch, 25, 26, 27, 28
  Posterior auricular, 12, 14
  Stylohyoid branch, 37
  Temporal branch, 12, 13, 14, 15, 16, 17
  Zygomatic branch, 15, 16, 17, 18, 19, 20, 21, 22, 23, 24, 28
Genitofemoral, 75
Glossopharyngeal, 109, 111, 112, 113, 116, 118
  Lingual branch, 107, 108
  Muscular branch, 117
Hemorrhoidal, inferior, (See Rectal, inferior)
Hypoglossal, 39, 40, 107, 108
  Descendens hypoglossi, 40, 41, 43
    Thyrohyoid branch, 42
  Muscular branch, 104, 105, 106
Iliohypogastric, 74, 76, 77
Ilioinguinal, 74, 76, 77
Intercostal, lower, 71, 73, 74, 76, 77
  Muscular branch, 65, 66, 67, 68, 69, 70
Laryngeal, external, 97, 114
    Recurrent, 98, 99, 100, 101, 102, 103, 114
Lingual, 107, 108
Lumbar, upper, 73, 79
Masseteric, 31
Oculomotor, inferior division, 91, 92, 94
  Superior division, 90, 91
Perineal, 89
  Branches of, 80
Pharyngeal plexus, 112, 114, 115, 116
Phrenic, 72
Pterygoid, lateral, 33
  Medial, 32
Pudendal, 82, 89
  Plexus, muscular branch, 81
  Perineal branches, 80, 82, 83, 84, 85, 86, 87, 88
Rectal, inferior branches, 80
Sacral, 4th, branches of, 80
Spinal, posterior rami, 53, 54, 55, 56, 57, 58, 59, 60
Stapedius, 96
Subcostal, 78, 79
Temporal anterior, 30
  Deep, 30
  Posterior, 30
Tensor tympani, 95
Thoracic, 12th, muscular branch, 78, 79
  Upper, 56
Trigeminal, 32, 36
  Mandibular division, branch, 110
Trochlear, 93
Vagus, 109, 111, 112, 113, 116, 118